MOTOROLA COLOR TV
SERVICE MANUAL
Volume Two

By Stan Prentiss

TAB BOOKS
Blue Ridge Summit, Pa. 17214

FIRST EDITION

FIRST PRINTING—FEBRUARY 1972

Copyright © 1972 by TAB BOOKS

Printed in the United States
of America

International Standard Book No. 0-8306-2584-4

Library of Congress Card Number: 76-91323

Preface

During the past three years or so, Motorola's solid-state plug-in board receivers have become so well known and respected that much of the television industry has followed this courageous lead which began with that stout "RF energy seeker"—the Original Quasar. Since then there have been many variations, especially with the emergence of integrated circuits. Also to consider are other Motorola achievements: the rectangular cathode ray tube, the new color-bright and now the black-surround tubes, tri-phase IC color demodulation (still a Motorola exclusive), "works in a drawer," Memomatic remote controls, Instamatic pushbutton brightness, contrast, and color correction, regulated AC and DC power supplies with short-circuit protection, plug-board exchanges, Drexel cabinets, and now a DC power source which is chopped at the horizontal sweep rate with excellent regulation—"you're dealing with a pro." We hope this second Motorola color manual, covering the solid-state and hybrid chassis produced during the past two years, reflects the company it keeps.

For those who helped with so much excellent information in the preparation of this manual, we have to thank Herb Bowden and Jim Smith, Sencore; Harry Rich & "company," Mercury; Jim Walcutt, Tektronix; Benton Scott, Motorola, Phoenix; Technical Services, Engineering, and Public Relations Departments, Motorola, Franklin Park; Henry Clarke and George Robinson, Washington Appliance Wholesalers, Motorola's Beltsville, Md. distributor; and John Novack, head technician at Service Kings, Kensington, Maryland.

Stan Prentiss

Contents

CHAPTER 1

The New Quasars

Motorola's 1972 line of color receivers includes a portable identified as the CTV7 (TS-931), console and table models, CTV8 (TS-938), and a remote-controlled model, CTV8 (ATS-938). These chassis are all solid-state—no tubes except the CRT. They have black-surround picture tubes, a marriage of the original Quasar and Quasar II plug-in panels, plus three new ones including, a simplified 3-function on, off, volume, channel change remote control.

A look at Fig. 1-1 represents the nine panels, some peripheral blocks and components, and a legend at the bottom right entitled "universal panel substitution." It says use CA for TA, and F for DA. Translated, this apparently means that, for instance, horizontal sweep panel F in the original Quasar can be substituted for horizontal sweep panel DA in the new Quasar, and color video panel CA in the Quasar II CTV6 can be swapped for the TA in the new Quasar. However, we understand substitution goes much further than this and that four panels in the Quasar II series and two in the original Quasar can all be used in the new Quasars. See Table 1-1. You would probably be well advised, however, to substitute only those panels that are completely up-to-date in any of the three types of receivers in order not to degrade performance. Actually, it is a good idea to turn in all panels once a year for the newest and most updated versions.

The three new panels consist of the YA remote panel on the ATS-938 chassis, the VA vertical and JA power supply panels. Apparently, differences between the TS-931 and the TS-938 series are in the cabinets and hardware. And since the CTV7 (TS-931) group includes portables, they do not have the much advertised "works in a drawer" feature. We'll discuss two of the three new panels, the VA vertical panel and the JA power supply, after considering some of the new receiver features as presented in Motorola's 1972 Technical Training Manual. The YA remote is covered in Chapter 6.

CTV8 CONSOLES & TABLE MODELS

Fig. 1-2 shows the locations of the front panel customer controls and Fig. 1-3 identifies the secondary controls located behind a flip-up door. The speaker panel is removed by pressing downward and pulling out, as shown in Fig. 1-4. To replace the speaker panel, insert the top edge first, then snap the bottom of the panel in position.

PORTABLE MODEL CTV7

The front panel control locations are shown in Fig. 1-5. The rear view in Fig. 1-6 reveals the locations of the Insta-Matic controls behind the back cover.

To release the drawer in those models using it, remove a ¼-inch hex-head screw near the AC, interlock. Press down on a clip at the lower guide rail and push the drawer forward at the antenna terminal block. A cheater cord can be inserted through a cut-out in the fibre-board back cover and connected to the interlock on the chassis.

An external speaker jack, located on the antenna terminal board (Fig. 1-8), allows sound to be fed to a speaker system. The jack is a standard phono plug. A pair of phono jacks (one on the TS-931) provides a monaural sound output suitable for application to a high-impedance amplifier input. Both 300- and 75-ohm VHF antenna inputs are provided. The UHF input is 300 ohms. The locations of the eight circuit panels in both receivers are shown in Fig. 1-9. Fig. 1-10 is a rear view of the receiver and power supply chassis. By removing four screws (A), the complete power supply can be removed or tipped up for service.

Two views of the power supply appear in Fig. 1-11, and the JA plug-in panel is pictured in Fig. 1-12. Fig. 1-13 is a view of the horizontal sweep panel and the new vertical panel is shown in Fig. 1-14. A convergence transformer supplies the proper signals to the convergence panel, HA (Fig. 1-15). Pincushion correction is accomplished by panel GA (Fig. 1-16). The HV adjustment control is located adjacent to the GA panel in the TS-931 chassis and on top of the drawer in the TS-938 (some sets) as shown in Fig. 1-15. To remove the front panel escutcheon, take out the speaker panel and four ¼-inch hex-head screws (A in Fig. 1-17). Remove two phillips-head screws (B) behind the secondary control door. All knobs must be removed.

To remove the secondary control bracket, take out four screws (A in Fig. 1-18), two at the top of the drawer and two at the bottom. The AFT and Insta-Matic switches can be freed from the bracket by

Panel	New Quasar	Quasar II	Original Quasar
AFT	KA	KA	---
IF & AUDIO	BA	BA	---
Col. Video	TA	CA, SA	---
Horz. Sweep	DA	----	F
Convergence	HA	HA	---
Pincushion	GA	----	G

Table 1-1. Interchangeable Quasar panels.

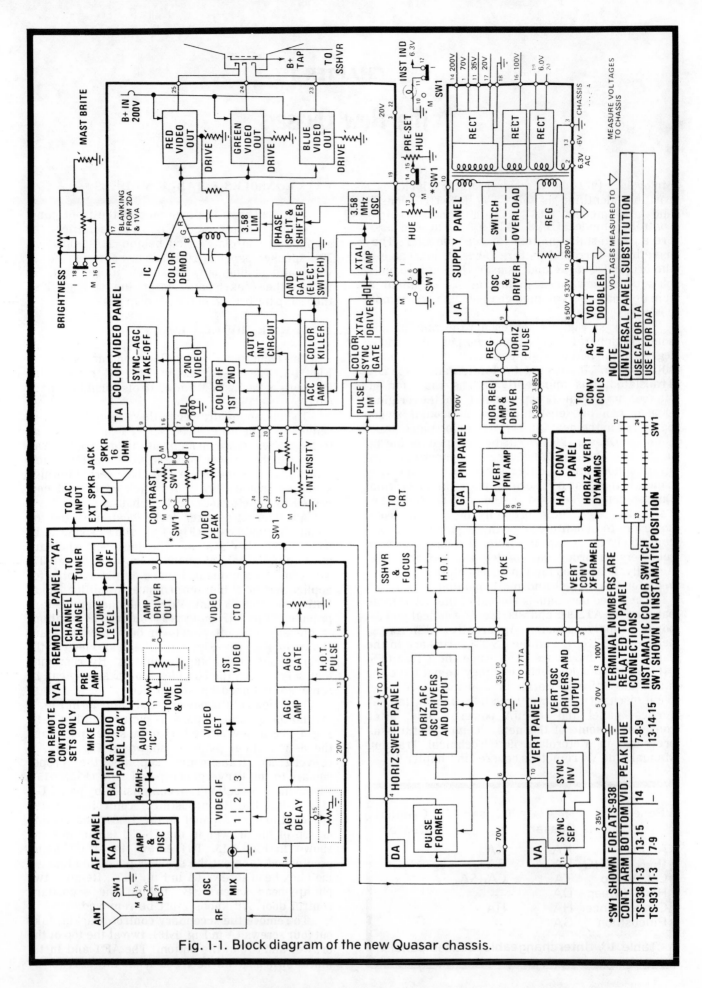

Fig. 1-1. Block diagram of the new Quasar chassis.

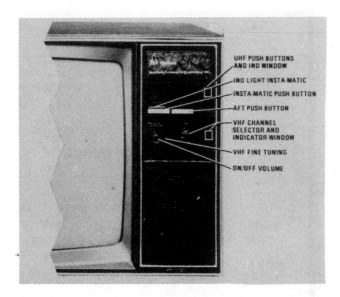

Fig. 1-2. Locations of the front panel customer controls.

removing two three-sixteenths hex head screws on each switch (B). The entire bracket should swing out. The VHF tuner is held by three ¼-inch hex-head screws. Be sure to disconnect all cables and plugs. To get to the UHF, remove the escutcheon and take out two ¼-inch screws (C in Fig. 1-18). Swing out the front control panel and remove two screws at the rear of the UHF tuner bracket (D in Fig. 1-19). Three ¼-inch screws secure the UHF tuner to a mounting bracket.

Setup, convergence and alignment instructions in Chapter 2 (Quasar II) are applicable to the new Quasars. Also, panels and circuits not covered here are identical to those found in the Quasar II.

THE VA VERTICAL SWEEP PANEL

The VA panel includes both sync separation circuits (Fig. 1-20) and the vertical oscillator and transformerless push-pull vertical output circuits.

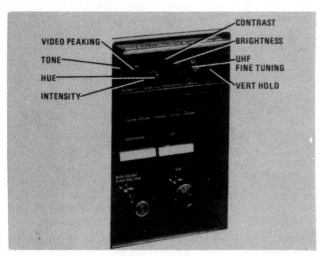

Fig. 1-3. Secondary customer controls are located behind a flip-up door on the front of the receiver.

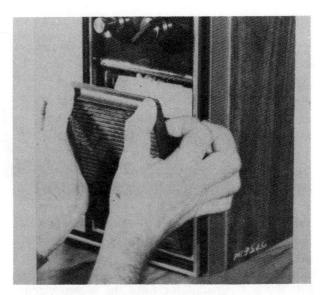

Fig. 1-4. The speaker panel is removed by pressing down and pulling out.

The vertical panel produces not only convergence waveforms, vertical deflection, and yoke drive, but vertical blanking as well.

Sync Circuits

Composite video with negative-going sync is applied to the base of Q5 (Fig. 1-2). The resultant positive-going sync at the collector of Q5 is applied to the integrating network, then diode coupled to the vertical oscillator. Horizontal sync is applied to the base of Q6 and appears at the Q6 collector.

Vertical Blocking Oscillator

Q7, T1 and the associated circuitry in Fig. 1-22 form a blocking oscillator. The frequency is determined by C15 and the vertical hold control. It is designed to operate at slightly below 60 Hz so that the incoming sync pulse will trigger the oscillator

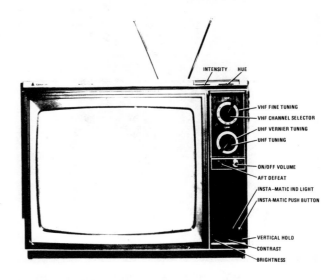

Fig. 1-5. Front view of the CTC7 portable.

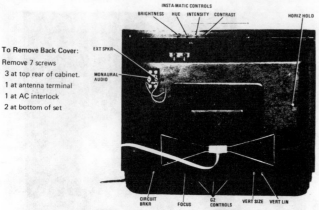

Fig. 1-6. Rear view of the CTV7 portable.

To Remove Back Cover:
Remove 7 screws
3 at top rear of cabinet.
1 at antenna terminal
1 at AC interlock
2 at bottom of set

just prior to the time it would be self-triggered. D1 provides coupling to the following stage. When Q7 is not conducting, D1 is reverse biased and provides isolation of the oscillator from the sweep circuits that follow.

Forming the Sweep Voltage

In the simplified circuit in Fig. 1-23, notice the saw-forming network (C5, R27 and the linearity control). During retrace time, Q7 is saturated and D1 is conducting. The positive pulse at the vertical output is applied to one end of C5. Current is flowing through Q7, D1, the linearity control and R27 to C5. Thus, a charge is placed on C5 with the polarity shown. When Q7 cuts off, D1 is reverse biased. Q4 conducts, but its base voltage rises slowly due to the charge on C5. Therefore, during scan time C5 partially discharges through R27, the linearity control and the height control. The resulting sawtooth voltage appears at the base of Q4. It is amplified and inverted in the collector circuit and appears at the output. During retrace, C5 charges again and the process is repeated. The linearity

Fig. 1-7. Front view with the "works in a drawer" pulled out.

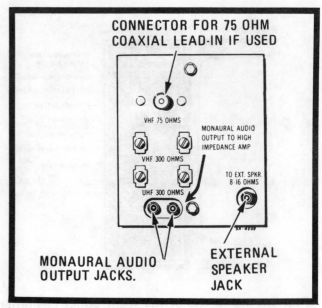

Fig. 1-8. Drawing of the antenna terminal board.

control alters the discharge rate of C5, thereby providing linearity correction. The height control varies the DC bias on Q4 to change vertical size. R29 allows the Q4 emitter to be operated slightly positive, allowing complete shut-off during retrace. C8 holds the emitter of AC ground.

Vertical Sweep Amplifier

The vertical sweep amplifier circuits (Fig. 1-24) are similar to the audio amplifier used in Quasar II receivers. Direct coupling eliminates the need for coupling capacitors between stages. The complementary symmetry output transistors provide a low-impedance output which is direct-coupled to the yoke. C1 prevents DC flow through the yoke, which would de-center the raster.

R11 and C9, from the Q2 collector and the predriver base, form a parabola voltage to modify the slightly nonlinear discharge of C5. C2, C3 and C7 suppress transients that could cause self oscillation. Vertical blanking pulses are formed from the retrace pulses that appear at the Q3 emitter.

SWITCH-MODE POWER SUPPLY

The block diagram in Fig. 1-25 shows the three basic functions of the new switch-mode power supply. First, the input AC voltage is rectified in a doubler circuit and provides a DC source for the following switch section. In operation this rectifier-doubler circuit is essentially the same as those found in most electronic equipment. In the switch section, DC is "interrupted" through the primary of a transformer in the same manner as in a conventional vibrator supply. Here, the transformer is considerably different and resembles a horizontal output transformer. In fact, this supply operates at the horizontal scan frequency and is synchronized at that frequency by pulses from the horizontal sweep circuit. The switch circuit is regulated to respond to changes in load and in this manner

Fig. 1-9. Circuit panel locations in both receivers.

provides regulated voltages to all receiver circuits. There is also protection from overload currents. The third function of the supply is to convert the input to suitable secondary voltages, rectify, filter, etc. Again the circuit are conventional.

Quick-on is provided by a separate line-operated transformer which is normally on when the set is off. Low power is supplied to the CRT filaments. When the set is turned on, the transformer is removed from the circuit.

Voltage-Doubler

A conventional voltage-doubler circuit provides DC voltages to the switch-mode oscillator, regulators, drivers and output circuits. Although a polarized line cord is used, the chassis is ground for all voltage measurements in the receiver other than in the doubler and primary circuits of the switch-mode transformer.

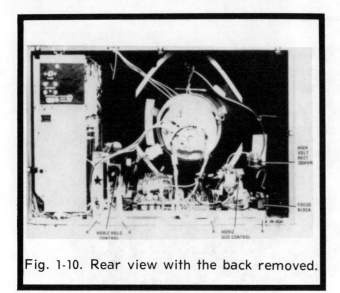

Fig. 1-10. Rear view with the back removed.

Most of these circuits are located on the JA panel. A 5-ohm 20-watt resistor provides protection for doubler diodes D800 and D801 from current surges due to charging the large capacitances in the circuit. Each diode is shunted by a .001-mfd capacitor to suppress radiation. The diodes are mounted on a plug-in board for easy removal. Oscillator and regulator circuits operate from the 33-volt zener regulated source at Terminal 6 on the JA panel.

The JA Panel

Q5 in Fig. 1-27 is a blocking oscillator operating at approximately 15,750 Hz and is synchronized by

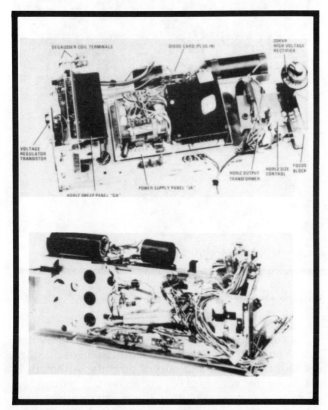

Fig. 1-11. Two views of the power supply chassis.

Fig. 1-12. Power supply plug-in panel JA.

pulses from the horizontal output transformer. T2 provides coupling from the output to the circuit in phase to sustain oscillation. The oscillator frequency is primarily determined by R15 and C6.

C7 in Fig. 1-28 charges rapidly through Q5 and pre-driver Q3 during horizontal retrace time. C7 discharges through regulator Q4, which acts as a variable resistor and controls the discharge time of C7 in accordance with the amount of forward bias applied. R19 adjust the Q4 forward bias and is the output voltage adjustment. D6 couples the waveform to the Q3 base.

When the regulator conducts very little or is nonconducting, its resistance is very high, resulting in a long time constant for the discharge of C7. With the regulator conducting heavily, the resistance is low. The result is a short discharge time, which allows Q3 to conduct early. By varying the discharge time, Q3 is made to conduct earlier or later in the period (Fig. 1-29). The square wave appearing at the Q3 collector, therefore, has a wide or narrow negative pulse, depending on when Q3 conducts. The pre-driver amplifies the negative pulse and R7 and R9 divide the output and direct couple it to Q2 (Fig. 1-30).

Drives Q2 and output Q800 alternately switch off and on. When the driver is conducting, the output is

Fig. 1-13. Horizontal sweep panel.

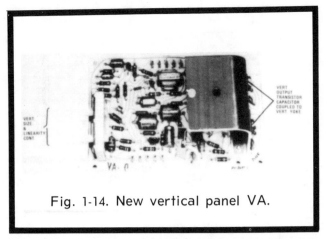

Fig. 1-14. New vertical panel VA.

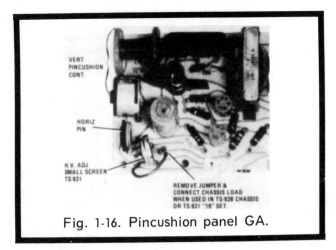

Fig. 1-16. Pincushion panel GA.

not, and vice versa. T1 acts as the Q2 collector load. It obsorbs the current energy flowing through Q2, releases it into its secondary, and drives output Q800. A square wave of voltage with a linear change in current is applied to the base of Q800. The collector load for Q800 is the primary of power supply transformer T802. Q800 acts like a switch, alternately completing the circuit from the 280-volt DC source and the T802 primary. The time duration (width of the driving pulse) that the switch is closed determines the amount of current stored (as a magnetic field) in the transformer.

The transformer secondary circuit consists of six separate windings and one tap (Fig. 1-31). Each source has a separate diode rectifier and filter. The CRT filament winding is also connected to a voltage divider (R20 and R23) from the 200-volt source to raise the filament to a potential close to that of the cathode.

An overload or short will automatically shut the system down. An abrupt change in the current requirements of the output transistor (Q800) results in a sudden voltage change which is applied to the gate of an SCR via a zener diode. The SCR is across Q2 and T1. When the SCR fires, the drive voltage is shorted to isolated ground and the supply shuts down. To reset the system, it is necessary to merely push the receiver on-off switch to off; then, after about 6 seconds, return it to on. The voltage across R4 in the Q800 emitter (Fig. 1-32) reflects the demands of the load.

Automatic regulation is accomplished by another winding on T802, which is tightly coupled to the 100-volt winding. The additional winding senses any changes in the load as changes in the voltage applied to D15, R18, R12 and R19. With an increase in the load, less positive voltage is developed across R18 and R19. The corresponding increase in the regulator (Q4) conduction widens the pulse (shorter time constant). The wider pulse allows the switch to conduct longer, resulting in more stored energy in T802 and an increase in the output.

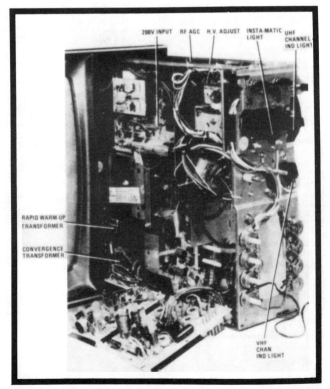

Fig. 1-15. View of the "drawer" with convergence and pincushion panels dropped down.

Fig. 1-17. Another view (partial) of the "drawer."

Fig. 1-18. Secondary control bracket in the service position.

QUASAR INTEGRATED CIRCUITS

At the 1971 IEE Spring Conference on Broadcast and Television receivers in Chicago, a brand new virtually all-IC (integrated circuit) color receiver was introduced by the Motorola Group from Phoenix, headed by Benton Scott. In it they have 10 integrated circuits plus an all-IC regulator (not discussed). Five of the 10 (or their virtual equivalents, except the color demodulator), are included in the new Quasars and more will be added as need and economics dictate. The ICs in the Phoenix receiver are: MC1364, AFT; MC1350, IF; MC1345, video processor; MC1398, chroma processor; MC1326, demodulator; MC1358, sound IF; MC1316, audio amplifier; SC1390, vertical deflection; SC1391, AFC and horizontal oscillator, and the MFC4060 regulator. The ICs we'll discuss are the MFC9020, audio; MC1352, video IF, AGC,

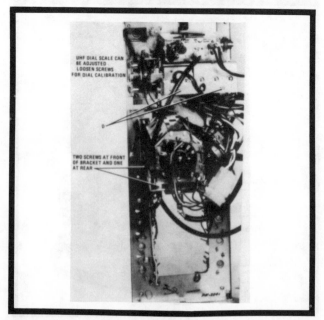

Fig. 1-19. Rear view of the "drawer" panel showing tuner mounting.

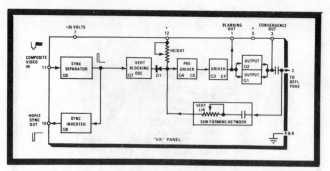

Fig. 1-20. Block diagram of the vertical panel VA.

and noise circuit; MC1398, color processor; MC1351, sound, and a slightly more sophisticated MC1358 chip. The tri-phase chroma demodulator developed by Receiver Engineering in Franklin Park, Ill., is fully discussed in the chapter on Quasar II receivers where it was **first** used. A block diagram of the Phoenix receiver and its 10 integrated circuits (in dotted rectangles) is shown in Fig. 1-33. Only the low-level video stages, video chroma outputs, horizontal driver and output-damper, and the tuners are not monolithic devices.

IC Video Block

In Fig. 1-34 are two stages of video IF, along with keyed AGC and an amplifier having a typical power gain of 52 db at the center IF frequencies, almost constant in and out admittances (Y equals 1/Z), two AGC potentiometers, two transformers and LC filters (not shown) for the sound carriers, adjacent-sound carriers, and upper-adjacent picture carrier. T1, as you can see, is double-tuned. L1 and the 33-pf capacitor between Terminald 3 and 2 of the MC1330 form a parallel tuned circuit inserted to prevent spurious operation of carrier-excited switches.

MC1352 or MC1353 IF, AGC & Keyer

These chips each have an AGC section operated by a composite video signal, flyback gating pulses,

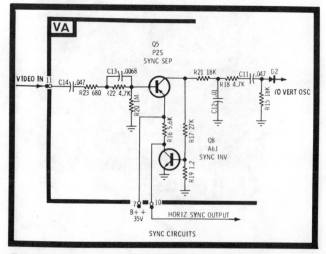

Fig. 1-21. Schematic of the sync circuits in the VA panel.

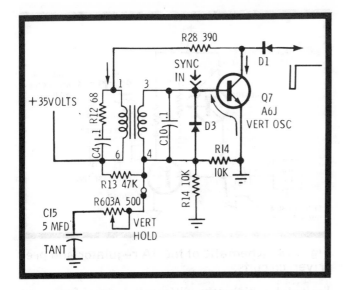

Fig. 1-22. Schematic of the vertical blocking oscillator used on the VA panel.

and reference level (Fig. 1-35). Either polarity video may be used, with positive signals applied to pin 10 and the AGC reference to pin 6. Flyback keying pulses come in through Terminal 5 to the base of the input amplifier, part of a differential, collectors-common configuration, with a current source at the emitters, biased like the other two current sources from the voltage dividers and a diode. The input zener is both bias and clamp for the input differential amplifier.

The IF input in the AGC controlled section is applied through pins 1 and 2 to the bases of Q4 and Q5 which are biased and regulated by the divider action of resistors and diodes from the V+ DC supply, just as are other bases and the collectors of Q8 and Q9. In Q4 and Q5, the input impedance is independent of the AGC, and input signals may either be single-ended or dual for differential action. Outputs from Q8 and Q9 are controlled by the magnitude of the IF AGC line on the bases of Q6 and Q7, and the resulting voltages go to the IF output (actually, the 2nd IF) supplied by a Darlingtonized constant-current source for steady quiescent bias. The MC1353 works with a negative RF-AGC voltage, while the MC1352 requires a positive RF-AGC.

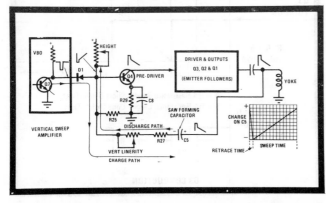

Fig. 1-23. Simplified circuit illustrating operation of the vertical deflection circuits.

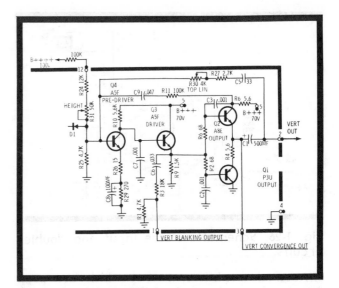

Fig. 1-24. Schematic of the vertical pre-driver and output stages.

The MC1330P Third Video IF & Low-Level Detector

The third IF is rather routine, but the detector circuit in Fig. 1-36 is most interesting. Very simply, Q7 is a constant-current generator, Q1 and Q2 are halves of a differential amplifier with a modulated carrier into Q1 and Q2 at AC ground. Q3, Q5, Q4 and Q6, the carrier excited switches, are in-phase and out-phase carrier inputs.

The output voltage across R1 is now detected modulation at twice the rate of the original 45-MHz carrier frequency. Stages that follow the detector have limited response and only amplify lower frequencies, making the detector self-filtering so that detector shielding against IF radiation is unnecessary. (What you really have here is a combination of analog input and digital switching that produces an ultra-clean video output at constantly controlled harmonic and amplitude levels.

AFC MC1364

The MC1364 automatic frequency control (or automatic fine tuning control-AFT) is a series of emitter-to-base coupled amplifiers (not illustrated), with zener supply regulators and three pairs of

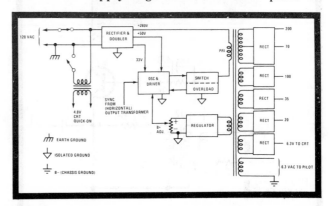

Fig. 1-25. Block diagram of the switch-mode power supply.

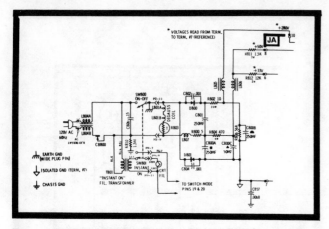

Fig. 1-26. Schematic of the input and doubler circuits.

differential amplifiers, two of which are peak detectors to supply a pair of differential outputs. A typical circuit application is shown in Fig. 1-37, with the incoming video carrier at 45.75 MHz. L1 acts as the discriminator primary, L3 as the second, and L2 as the tertiary (3rd) winding. What this circuit does is to sense frequency deviation on either side of the video carrier through the discriminator transformer and deliver a plus or minus output to the AFC to return the tuner oscillator to its proper frequency so that the AFC again sees exactly 45.75 MHz at its input.

The MC1358 Sound IF Amplifier

The sound IF amplifier is interesting enough to show the entire schematic in its six sections, including an audio amplifier, IF amplifier limiter, a regulated power supply, an electronic attenuator, a detector, and an output emitter-follower buffer (Fig. 1-38). What happens is that sound comes through a 4.5-MHz selective transformer into the sound input (1 and 2); it is amplified and precisely

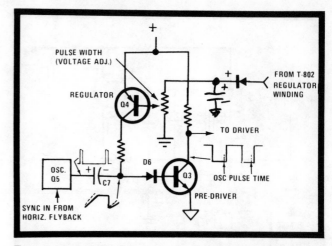

Fig. 1-28. Schematic of the JA regulator and pre-driver circuits.

limited by high-gain differential amplifiers and passed through the somewhat "Darlingtonized" stages into the audio detector. There, the transistor emitters and 15-pf capacitors act as peak detectors. The resulting signal is applied to the emitters of the center differential amplifier in the electronic attenuator, where bias for these circuits is derived from a transistor in the regulated power supply. The volume control is connected to pin 6. Pin 7 goes to a de-emphasis capacitor and pin 8 connects to the sound input, pin 14. The audio output signal appears at pin 12.

MFC9020 Audio Output

The audio output IC (Fig. 1-39) is a 2-watt amplifier with heat sink. Basically, it is another differential amplifier with Darlington current-driver outputs (not shown) with a low enough output impedance to drive a capacitively coupled 16-ohm

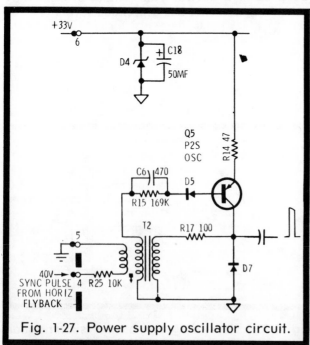

Fig. 1-27. Power supply oscillator circuit.

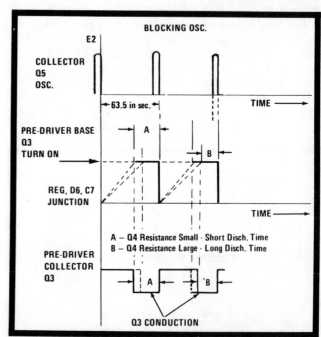

Fig. 1-29. Waveforms illustrating operation of the oscillator and pre-driver.

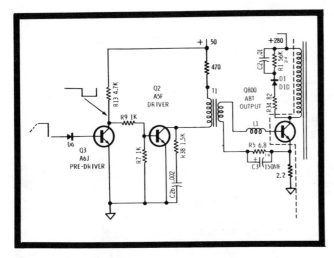

Fig. 1-30. Schematic of the driver and output stages.

speaker through a 250-mfd capacitor, thus eliminating a step-down transformer.

MC1345P Signal Processor

The IF circuits in Fig. 1-34 are all contained in the TV signal processor, consisting of a sync separator, advanced noise inverter, AGC comparator, and RF AGC delay amplifier (Fig. 1-40). The noise threshold can be manually adjusted, the sync separator time constants preferentially set,

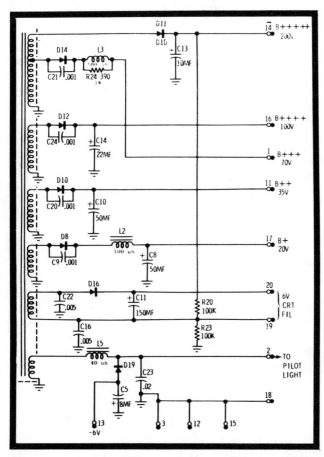

Fig. 1-31. Schematic of the transformer secondary circuits.

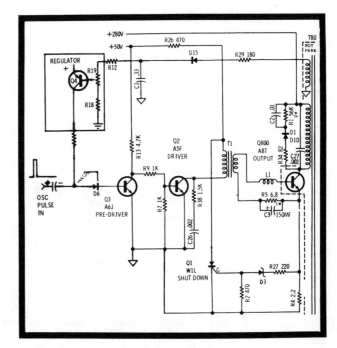

Fig. 1-32. Schematic with the regulator stage added.

and the chip regulated for 10 percent power supply variations.

Composite video with negative-going sync enters the video input at pin 1, riding on a positive DC forward bias. Bias from the emitter of Q32 and any external resistance connected to the emitter of Q7 determines the noise threshold. Signals at the Q6 emitter appear across the Z1 through Z4 bias chain, in which the anodes are connected to 2.6 volts supplied by Q5. The zeners and dropping resistors constitute an RC delay line of 300 nanoseconds. The video signal, inverted by Q10, is current reinforced through Q11 and Q12 to the output, pin 13. Should noise appear, the emitter of Q6 is driven below its pre-set level, Q7 conducts, turning on both Q8 and Q9 and grounding the collector of Q10 and the base of Q11, thus blanking the output signal. Blanking begins before the noise pulse completely passes through the delay line and remains for a short period after the noise has passed, due, probably, to the reinforcing charge stored in the 0.1-mfd capacitor connected to the emitter of Q7. If allowed to fully discharge, this capacitor will cancel output signals for four lines of horizontal sweep.

Positive-going, noise-free but filtered video now enters the base of sync amplifier Q35 through input pin 12, which is not (as shown) connected to the 18-volt supply as the emitter of Q36 should be. Sync tips cause conduction of only the sync chain and produce a negative-going, low-impedance 15-volt output from the Q38-Q39 complementary pair.

Delayed video into the AGC keyer originates from the emitter of delay amplifier Q10 and is connected to the base of Q17, while sync pulses go to AGC keyer Q13 and Q14 from the collector of Q36 in the sync separator. AGC keying from the flyback transformer is applied at pin 9 to the base of Q15, then Q13 through Q15 operate together, as flyback and sync pulse tips coincide to produce the AGC

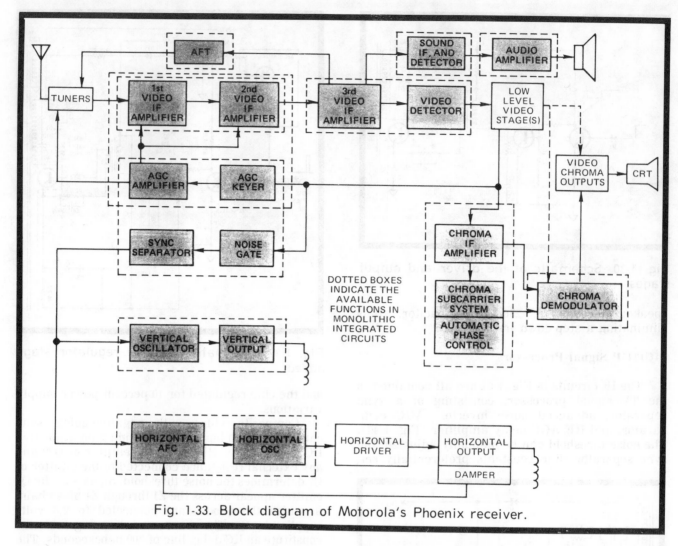

Fig. 1-33. Block diagram of Motorola's Phoenix receiver.

keying action. The succeeding stage is an AGC comparator, with signal from Q14-Q15 reaching Q17 through the base-collector of gated current source Q19, while the base of Q18 is fixed by the 2.6-volt bus. Q19 conducts only when Q14 and Q15 are switched off by the coincidence of a positive sync pulse on the base of Q13 and a negative flyback pulse on the base of Q15, thereby permitting forward base bias for Q18 from the positive power supply. Negative sync pulses from Q10 now pass through the base and emitter of Q17 into the emitter of Q18, forward biasing this transistor which, with no phase inversion, turns on Q20 and Q21 to charge a 2-mfd capacitor connected to pin 8. Should the video signal and negative sync pulses decrease towards zero, Q18 will not conduct, but switch Q22 will conduct, decreasing the stored charge on the 2-mfd capacitor and allowing rapid AGC changes because of the charge-discharge transistor-capacitor action. Q19, with no collector load or reverse bias, would be cut off.

The AGC keyer output passes through the base-emitter of Q40 to the bases of Q23, then Q26, and part of a differential amplifier with Q27. Q4 is the common-emitter current source for both. On weak signals, when Q26 is barely conducting, Q27 can be biased on by a DC RF delay potentiometer at pin 6. The Q27 collector voltage drops, turning on Q30,

Q31, and Q24, which furnishes both negative RF AGC through Q31 (pin 2), the collector supply for Q3 and the base bias for Q25. Thus, Q25 turns on somewhat harder. When Q40's current increases, Q26 begins to come on, turning on Q28 and Q29 for a positive RF AGC output and turning down or off Q30, Q31 and Q24, thus reducing or eliminating the negative RF AGC and providing less base bias to Q25.

Vertical Oscillator & Output, XC1390P

In this virtually all-IC Motorola receiver, there is a monolithic vertical oscillator and output. The age-old vertical linearity control is eliminated, there are independent size and vertical hold controls, scan current is independent of deflection yoke variations, and the output is short-circuit protected. Fig. 1-41 is a typical output circuit with filtering, bias and signal inputs, plus the various controls. Notice that there is no vertical output transformer; simply AC coupling through a 250-mfd capacitor to the deflection yoke and convergence circuits. A vertical "flyback" stage is included so that during retrace a pulse greater than the DC supply simulates the usual high peak voltage of the vertical output transformer. This capacitor has a value of

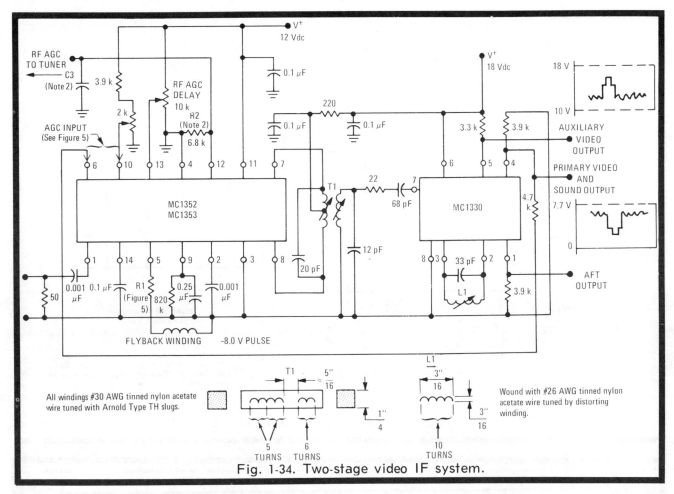

Fig. 1-34. Two-stage video IF system.

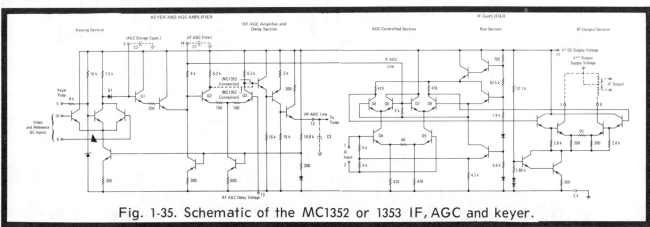

Fig. 1-35. Schematic of the MC1352 or 1353 IF, AGC and keyer.

150 mfd and it charges through external diode D1 to the power supply.

The **oscillator** (Fig. 1-42) is "timed" by a 0.33-mfd capacitor that, during forward scan, is charged through transistor Q25 until Darlingtonized differential amplifier Q34-Q35 is switched by a negative sync pulse entering the IC through pin 2. As the collector of Q35 goes low, Q38 switches on, latching Q37, which turns on switch Q6 and turns off Q5 and Q7. Q3, now forward biased, conducts and discharges the 0.33-mfd microfarad capacitor as differential amplifiers Q30 and Q31 commence conduction, turning on Q2 but turning off Q37 and

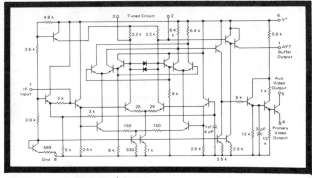

Fig. 1-36. Schematic of the MC1330 third IF and low-level video detector.

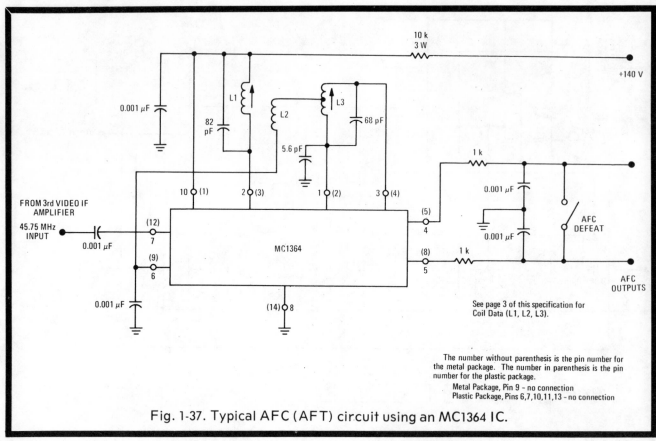

Fig. 1-37. Typical AFC (AFT) circuit using an MC1364 IC.

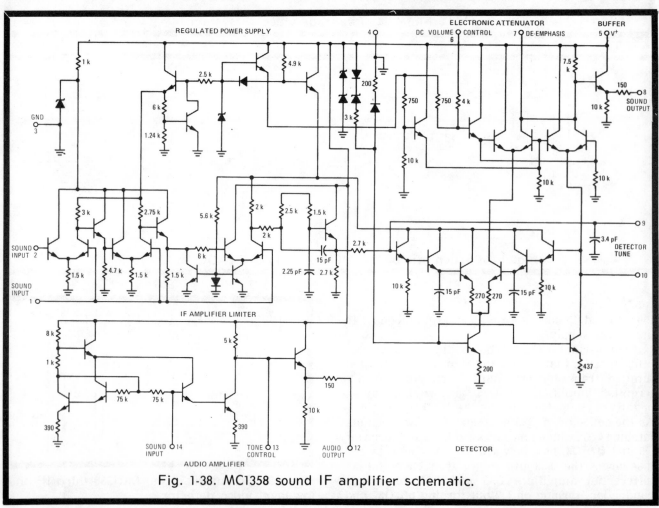

Fig. 1-38. MC1358 sound IF amplifier schematic.

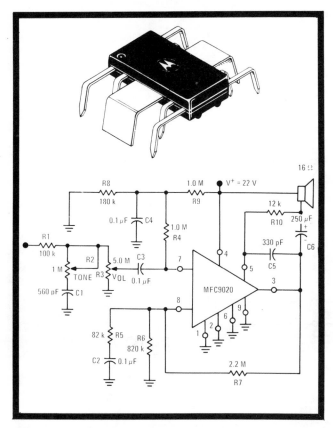

Fig. 1-39. Audio output using an MFC9020 IC.

to the power supply. When Q8 and Q9 are turned off, and switches Q15 and Q19 clamp the bases of Q16, Q17 and Q18 to ground, current flows from the yoke into output pin 8, through D3, out of pin 9, into Cf, through flyback pin 12, D4, and through the base-collector junction of Q23 to the V+ power supply (pin 13). As the coil's magnetic field reverses, Q23 and Q24 become diodes to clamp the negative side of Cf to the supply voltage and diode outputs Q21 and Q22 saturate with output bias. As retrace commences, external D1 is reverse biased and the capacitor delivers a flyback voltage larger than the supply voltage, forming the conventional retrace spike.

The output stages are driven by the oscillator sawtooth coming from the ramp output through a 20-mfd capacitor into pin 7 and bias is adjusted by the height control. Quiescent voltage is developed through series divider R14, R15, R16, and Q11-Q12 Darlington, and another series divider R17, R18 in the emitter of Q12. Q13, Q14 and Q17 supply this potential to the base of Q18 where it is developed across R24 for the Class A quiescent current, along with further set and ripple filtering applied by a parallel RC combination connected to pin 1. Q10 switches on to limit the Class A current and so limits dissipation at high temperatures.

During forward scan, Q17 and Q18 turn on slowly, removing current at the output. Q16 starts to conduct, dropping voltage across R29 and removing the potential from R26, R27, R28, reducing the emitter bias of Q21 and Q22, allowing less current output. As currents decrease in Q21 and Q22, they increase in Q17 and Q18, until at the center of the scan the currents equal the Class A quiescent current and the output current stops. As the forward scan continues, Q17 and Q18 continue delivering

Q38. The sawtooth output of Q3 is now coupled through Darlington stages Q39 and Q40 through R1, the ramp output at pin 6.

During forward scan, 150-mfd capacitor Cf is charged by current sources Q8 and Q9 through D4. Terminal 12, and external diode D1, which connects

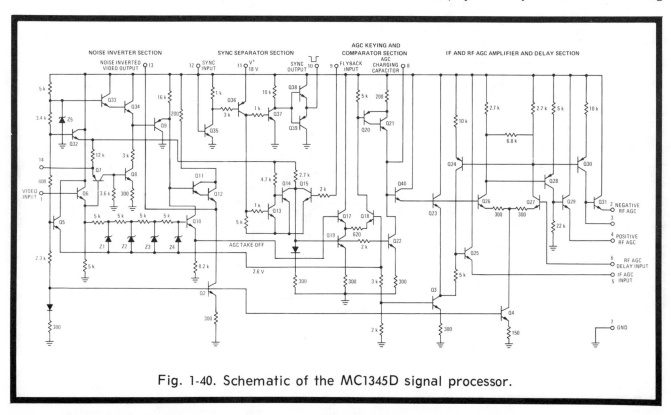

Fig. 1-40. Schematic of the MC1345D signal processor.

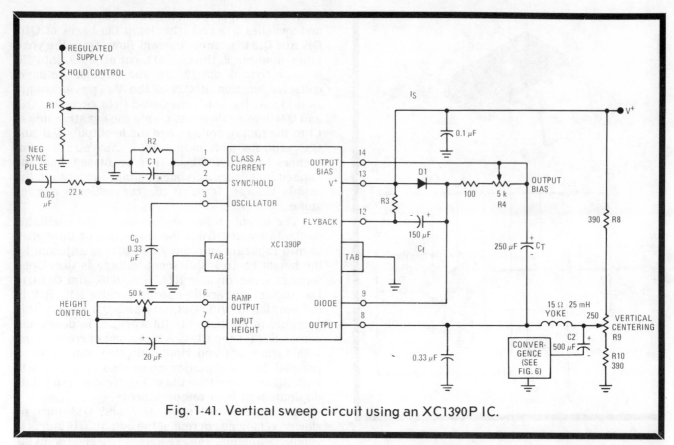

Fig. 1-41. Vertical sweep circuit using an XC1390P IC.

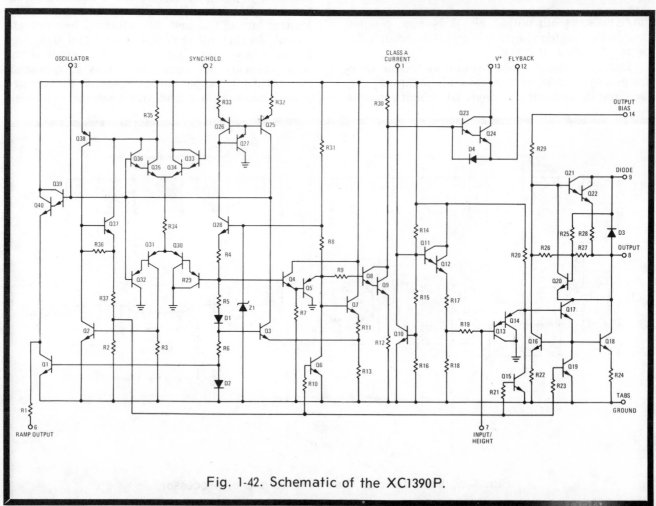

Fig. 1-42. Schematic of the XC1390P.

additional current until, at the end of the scan, these two transistors are taking away peak current due to voltage across R24, while Q21 and Q22 are virtually cut off due to heavy current flow in Q16. The circuit is now ready for retrace.

The Color Processors & Demodulators

The two final intricate monolithic chips in this Motorola IC receiver are the color processing circuits and the chroma demodulator-drivers. A typical applications circuit appears in Fig. 1-43. You see one input (T1) and one interstage (T2) transformer, with a pair of trouble coils. Other components, except the crystal tuning capacitor and a couple of DC controls, are fixed. Notice that both blanking and luminance are put into the second MC1326 chip, with all analog output signals DC coupled to discrete CRT drivers and the cathodes of the picture tube. So far, Motorola is the only manufacturer we know of who has an RGB system output from its MC1326 demodulators. The remainder have (RGB)-Y, with mixing taking place in the final output stages or even the cathode ray tube.

As shown in Fig. 1-44, the MC1398 color processor is divided into nine sections consisting broadly of color amplifiers, burst gating, color killer, phase shifter, 3.58-MHz subcarrier oscillator, automatic chroma control, and voltage regulator, with a bit more in between. The chroma input is applied at pin 5 through 3.08 to 4.08-MHz frequency-selective input transformer T1 to the emitters of differential amplifiers Q3 and Q9, with the output

appearing the collector of Q3. Bias for both Q2 and Q3 originates from the diode divider string, D1 through D8, as well as for other transistors connected higher on the line to the 24-volt Vcc supply. On the lower right, the second ACC "filter" circuit controls the gain of Q19 and Q9, while Q6 (upper left) is a regulating source for the collectors of Q5, Q7, Q3, and the base of Q8. The ACC-killer control at pin 10 determines the level at which Q46 and Q47 supply base bias for Q19, and for Q24 and Q25, a Schmitt trigger with an input transistor that is on when its output is off or vice versa. Q25 must be in saturation for the color killer to cut off chroma action when there are either noisy color signals or monochrome only. Therefore, a "high" input from Q46 and Q47 produces more B+ bias for Q19, turning it on and turning off Q25, allowing full color passage.

As chroma amplifier Q3 operates, Q8 passes signal to Q17 and its Q18 current source to the emitters of Q12-Q16. Here, chroma is gated by a 4-microsecond horizontal pulse through differential amplifiers Q5 and Q7, aided by level control amplifier Q22 and Q26. Further amplification and control is provided by current-sharing amplifiers Q11 and Q15, with chroma output through Darlington stages Q14 and Q20. The burst gating and fill-in circuit also sends the broadcast color sync burst to the base of Q27 (in the oscillator section) from the collectors of Q26 and Q16 to lock the 3.58-MHz subcarrier regenerator to the transmitted color sync frequency.

A 3.579545-MHz crystal with a 5 to 20 pf trimmer normally controls the subcarrier oscillator which is always subject to close tolerance burst correction.

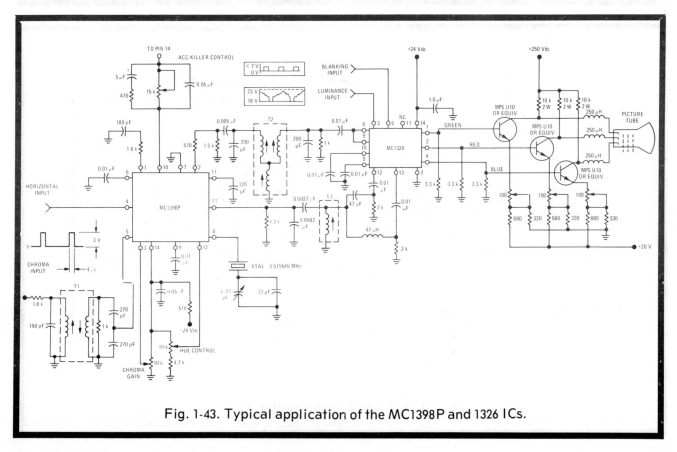

Fig. 1-43. Typical application of the MC1398P and 1326 ICs.

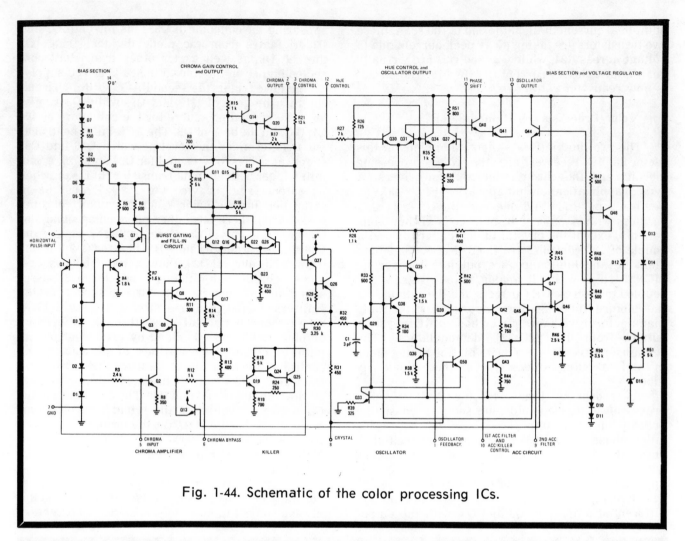

Fig. 1-44. Schematic of the color processing ICs.

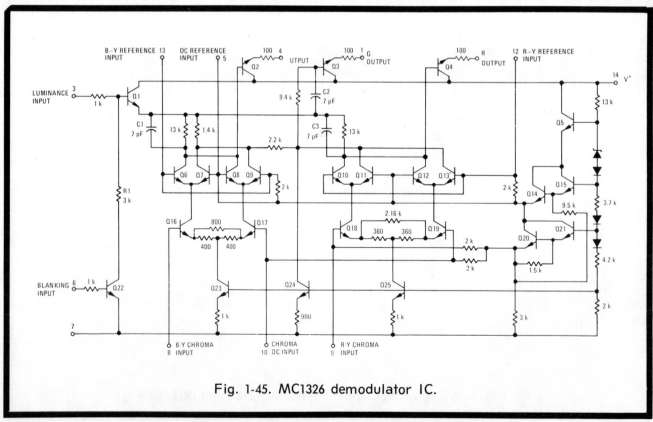

Fig. 1-45. MC1326 demodulator IC.

Differential amplifier Q29 and Q50 receives the sync information, with feedback from Q35 and the RC series circuit at pin 1, while Q38 and Q39 provide the hue control and oscillator output (pin 13) with a pair of 180-degree out-of-phase signals. The hue control is another DC positive voltage that changes bias mainly on Q30 and Q37 and, therefore, the forward conduction of Q31 and Q34, "summing" the two differential amplifier outputs in the collector of Q37 in varying degrees and thereby changing the output phase angles of Darlington transistors Q40, Q41 and the signal at pin 13. The Q50 oscillator output is rectified by offset differential amplifier Q42 and Q45 to supply a DC bias voltage for Q13 and Q9 that is proportional to the input signal level. The bias section and voltage regulator are just what their names imply, and are DC references through Q44, Q48 and Q49, with zener reference D15, which insures internal regulation of the chip over large external voltage changes.

MC1326 dual doubly balanced chroma demodulator (Fig. 1-45) receives chroma at the differential pairs Q16, Q17 and Q18, Q19 through inputs at pins 8 and 9, while DC references at pins 10 and 5 are held at AC ground through 0.01-mfd capacitors. A 3.58-MHz subcarrier reference, phase shifted approximately 104 degrees by external RLC components, is applied to the R-Y and B-Y inputs at pins 12 and 13, respectively. R-Y oscillator references go directly to Q10 and Q13 and B-Y references proceed to Q6 and Q9. Chroma information from the lower differential pairs is switched by the upper differential pairs—Q6 through Q13—at the subcarrier frequency rate, with all transistor collectors being cross coupled for "doubly balanced" or full-wave synchronous detection. The negative outputs of these differential pairs are resistively matrixed to form the center green output.

With transistor Q1 (upper left) connected in the collector circuits of the synchronous demodulators, luminance information can be injected and mixed almost unattenuated at the collectors of these transistors and directly produce red, blue, and green outputs at Q2 through Q4. Capacitors C1 through C3 compensate for high-frequency rolloff and supply filtering for carrier harmonics. Dual horizontal and vertical blanking is introduced through switch Q22 that, when driven close to saturation, pulls down the base of Q1, causing its emitter to follow and so reduce the demodulator output to a non-CRT driving condition but not remove all supply voltage from the demodulator collectors, a condition that would upset chroma-oscillator lock after each blanking pulse.

CHAPTER 2

Quasar II Receivers

The Quasar II receivers are the CTV5 and CTV6 sets that differ little except for power supplies, "works in a drawer," mechanical cabinet features, and a modest number of changes in several external circuits. All plug-in panels are the same. Otherwise, the circuits for these two receivers are relatively straightforward and seem to need no additional explanation. Integrated circuits that may appear later in some of the Quasar II panels are described in Chapter 1.

TUNERS

The CTV5 use OPTT-429 or OPTT-430 (Fig. 2-1) VHF tuners, featuring off-set pre-set fine tuning. They are directly interchangeable. Both use all plug-in connections and varicap circuitry for automatic fine tuning. The CTV6 uses (concentric pre-set) tuners. CPTT-429 or CPTT-430 (Fig. 2-2) on models with AFT. Both tuners are interchangeable. Models without AFT use the CPT-425 (Fig. 2-3).

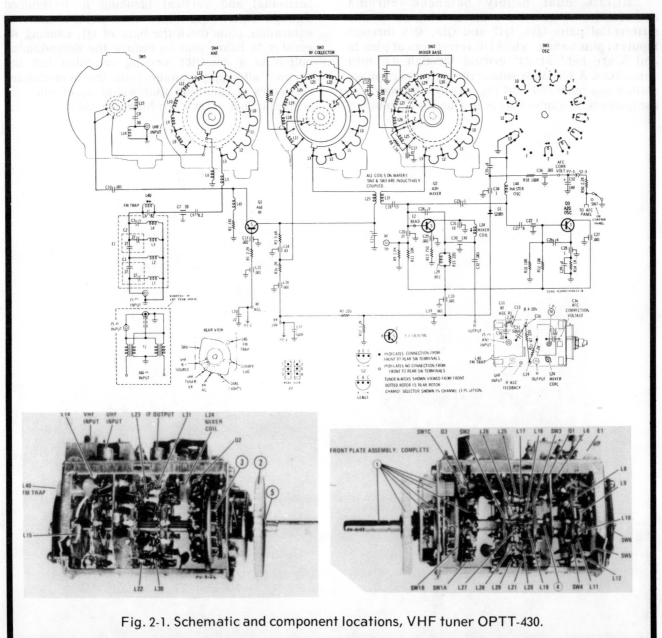

Fig. 2-1. Schematic and component locations, VHF tuner OPTT-430.

All the above are 4-circuit switch-type tuners using three transistors. They also use a grounded-base RF amplifier for improved cross-modulation characteristics. Also, inductive coupling is used between the RF and mixer. The oscillator is conventional.

The CTV5 uses the TT-645 UHF tuner (Fig. 2-4) and the CTV6 with AFT uses the TT-640 (Fig. 2-5) and non-AFT models the TT-650 (Fig. 2-6). All are similar to previous solid-state UHF tuners.

In AFT models, when a station is properly fine tuned the video carrier is at 45.75 MHz. Some of this signal (applied to the KA panel) is amplified in Q1 and Q2 (Fig. 2-7) and applied to a discriminator circuit which is tuned to 45.75 MHz. Under these conditions, the circuit is balanced and no voltage appears at the output, hence no correction would be applied to the tuner. If the fine tuning is now misadjusted, the video carrier would no longer be at 45.75 MHz. As a result the AFC circuit would be

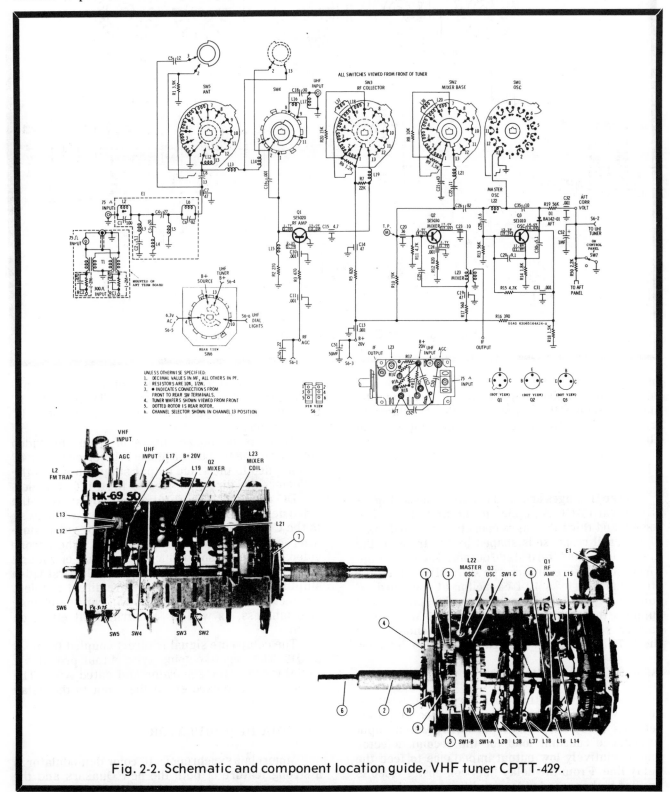

Fig. 2-2. Schematic and component location guide, VHF tuner CPTT-429.

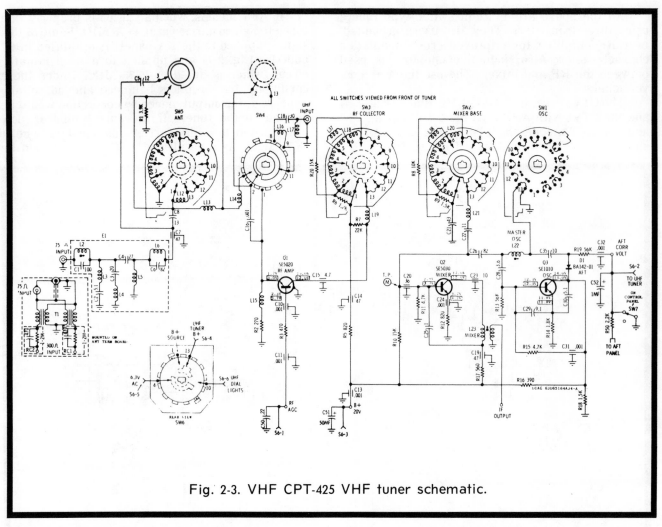

Fig. 2-3. VHF CPT-425 VHF tuner schematic.

unbalanced and would develop a voltage (plus or minus). This voltage is applied to the varicap in the tuner, changing the capacity in the oscillator to bring it back on frequency. For alignment, refer to Chapter 3.

VIDEO IFs

Three IF stages are used in both chassis (Fig. 2-8). Forward AGC is applied to the first stage. The second and third IF stages operate with fixed bias. The overall response is shaped by the traps in the first IF input and third stage output. The interstage tuned circuits are broad. The 45.75-MHz AFT take off is derived by a 1-pf capacitor at the third IF collector. A 2.7-pf capacitor at the same point supplies the audio detector. The 41.25-MHz audio carrier is trapped out in the third IF secondary, and the 45.75-MHz video is applied to the video detector.

B&W SIGNAL PATH

The signal out of the video detector has positive-going sync and blanking. The first video amplifier (emitter follower) provides a relatively high input impedance constant load for the second detector and a relatively low output impedance to feed the delay line. From the delay line, composite video is applied to the second video amplifier, operating as a

split load. Signal at the emitter of this stage is applied across the contrast control. Then to the IC through the secondary of the second color IF transformer.

There is no phase inversion from the video detector to the IC, but the phase is inverted by the IC and applied to the video output stages in which the collectors are direct coupled to the CRT cathodes.

DC coupling (Fig. 2-10) is employed from the video detector through to the IC (in fact all the way to the CRT). Forward bias for the first video amplifier is established by R21 and R39. The second video amplifier also conducts as a result of voltage applied to its base from the emitter of the first video amplifier. Because these stages are direct coupled, a defect in the video amplifiers can cause a change in brightness, as well as a detrimental effect on video.

The composite signal is direct coupled from Q3 to Q18. The negative-going sync output provides a signal for the sync separator and gated AGC. The color signal is picked off at the input to the delay line.

CHROMA DEMODULATOR

Motorola's synchronous IC color demodulator is a strong feature of both the new Quasars and the Quasar II series. The "trick" in obtaining the color

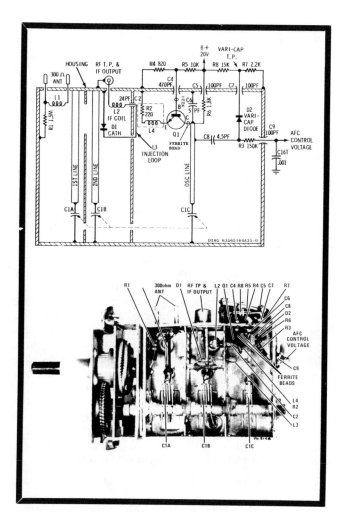

Fig. 2-4. UHF tuner TT-645 schematic and component locations.

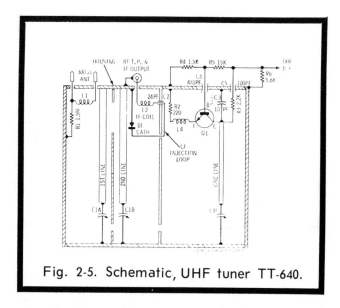

Fig. 2-5. Schematic, UHF tuner TT-640.

tertapped secondary of T1, and composite video is processed in 180-degree positive and negative phase opposition to the picture information inputs of all three chroma demodulators. Within this block diagram is also indicated a voltage regulator and a combination brightness and blanker function we'll discuss next.

rendition realized is due to full-wave processing in the balanced demodulators and to mixing the fine detail picture (luminance) information internally within the integrated circuit demodulator chip. The result is a completely controlled and processed intelligence which is delivered directly to the cathodes of the picture tube.

As you can observe in Fig. 2-11, there are three different signals used in this tri-phase demodulation process. The 3.579545 MHz regenerated subcarrier from Q14 is applied directly to the green demodulator, through an RLC network into the blue demodulator, and through an RC net to the red demodulator. Immediately, this tells you that natural green is the 3.58 MHz reference and the phase angles of the other two are attenuated so that the difference between the angles of red and blue demodulation should, with a yttrium red phosphor, amount to about 105 degrees—the usual industry standard. On a vectorscope, the green will appear (if keyed rainbow generated) at a relative phase angle of 300 degrees, while positive red is around 90 degrees and blue about 195 degrees. The keyed rainbow generator, of course, has burst reference at 0 or 360 degrees, and all other phase angles are referred to it.

Color information (chroma) is present at T1. Monochrome is mixed with chroma via the cen-

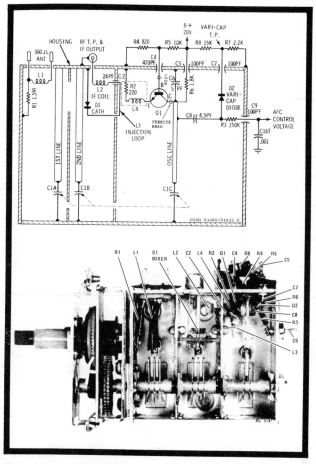

Fig. 2-6. UHF tuner TT-650 schematic and component locations.

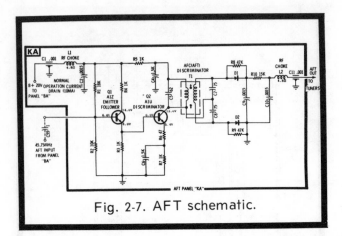

Fig. 2-7. AFT schematic.

Basic Demodulator, Blanker-Brightness Functions

To better understand demodulator operation, look at the single stage in Fig. 2-12. Chroma at phase differences of 180 degrees is applied to inputs 1 (Q5) and 2 (Q6). Both inputs are controlled by closely matched emitter resistors R3 and R4 for equal current output—a condition that is required to insure a 40 db rejection gate drive and keep 3.58-MHz subcarrier signals out of the demodulated output to prevent unnecessary transient trapping. R5 controls the differential gain and helps balance R3 and R4. With positive chroma and subcarrier coming into Q5 and Q1, respectively, Q5 conducts, Q1 conducts and current flows through R1. On the following negative oscillator input, Q1 shuts off, but because of static base bias Q2 conducts, forcing current through R2. Negative-phase chroma also operates amplifier Q6, and this transistor turns on and off, allowing Q3 and Q4 to conduct, alternately supplying R1 and R2 with initial or additional

current, whichever is the case. J1 adjusts the signal gain between the series plate load resistors. The two phases of chroma demodulation now result in full-wave recovery of all chroma information and, with luminance added in the collectors of these four transistors, RGB outputs are delivered to amplifying stages that directly drive the red, blue, and green cathodes of the picture tube.

In Fig. 2-12, you see a simplified outline of the blanking (Q25) and luminance (Q9) inputs. The first is a simple gate that reduces the demodulator supply voltage by some 6 volts during retrace. This is the time when vertical and horizontal retrace (we identify here as blanking functions) take place and no picture should show on the screen while both sweeps are recovering to start another trace. Between blanking intervals, luminance information comes in through Q9 and controls the B&W picture content and background illumination.

The entire demodulator schematic appears in Fig. 2-14. Each stage works exactly like the one described. Notice that the luminance input, Q26, controls the base of Darlington Q24 that, in turn, fires Q23, and this transistor then feeds the collectors of the three pairs of balanced demodulator switches with DC as well as AC. The load for Q26, however, goes right to B+ via the collector of Q25 (blanker). This means that when Q25 conducts, brightness is cut off and the entire DC level for emitter followers Q1, Q11, and Q20 is lowered, so that there is no picture output.

INSTAMATIC

Motorola's Instamatic employs the usual preset potentiometers for contrast (R109), (Fig. 2-15) brightness, color intensity, and hue—all initially

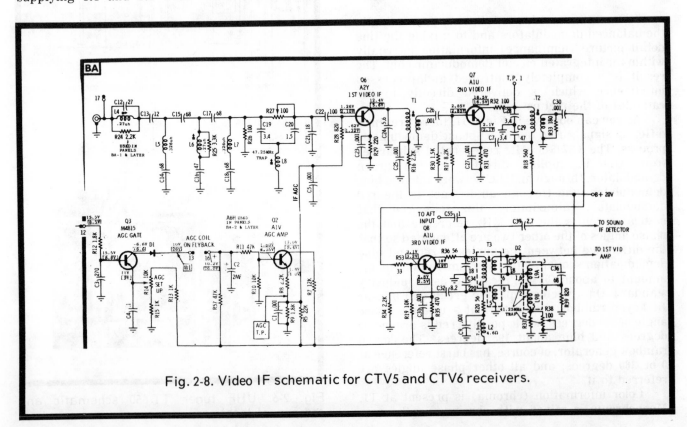

Fig. 2-8. Video IF schematic for CTV5 and CTV6 receivers.

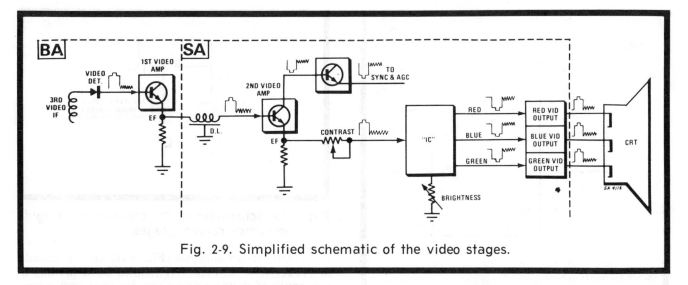

Fig. 2-9. Simplified schematic of the video stages.

adjusted by the manufacturer. But the most interesting portion of the Instamatic circuit really lies in the action and interaction among four transistors on a special PA color preset panel that is part of the CA color panel in the Quasar II receivers. And while Fig. 2-15 shows the preset controls and part of the actual schematic, more of the schematic is graphically portrayed in Figs. 2-16, 2-17, where other related circuits are also included.

When the front panel switch is depressed, more red is added to the picture and the angle of demodulation is opened, broadening the general area of fleshtones. At the same time, a closed-loop circuit automatically supplies a separate control voltage to the second color IF amplifier, thus limiting the voltage gain of both the first and second color IF amplifiers. The result is an automatic circuit that prevents unpleasant color overload. And since the brightness, contrast, and AFT (automatic fine tuning) are already switched and operating, the receiver is completely tuned from the oscillator in the tuner, to screen illumination, gray scale, hue,

and chroma amplitude—a complete video tuning operation that is wholly automatic in all respects.

To understand the operation of the Instamatic unit, you'll almost have to look at Figs. 2-16 and 2-17 alternately. Both circuits are dependent, and the color killer cannot operate without the two internal amplifiers in Fig. 2-17, while the two internal amplifiers are dependent on incoming color.

When color arrives at the receiver, the automatic chroma control amplifier turns on, biasing on the color killer, shown as Q11 in Fig. 2-16. The collector voltage for this stage comes from terminal 15 and the 120K (R8) resistor from the second internal amplifier. The second internal amplifier output is rectified negatively by a series diode and goes through zener DIU to Terminal 15, in the automatic position (Fig. 2-17), and then to the base of the second color IF, or, in manual position, to the collector of Q11, and through R51 and R900, the intensity control and finally again to the base of the second color IF. Of course, when there is no color, the base of the color killer is biased positive

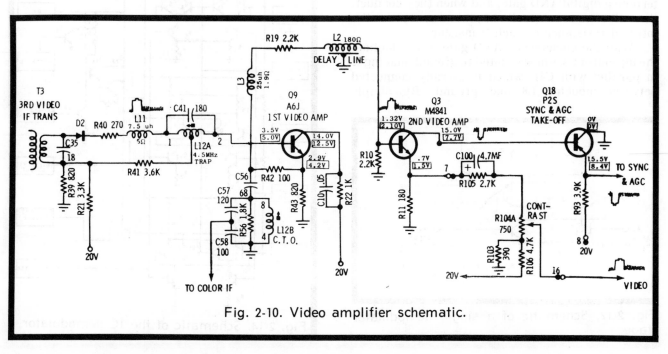

Fig. 2-10. Video amplifier schematic.

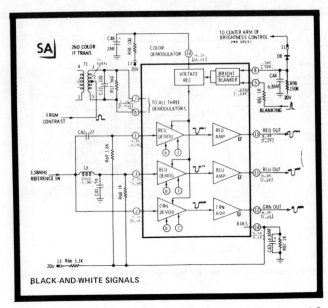

Fig. 2-11. Simplified schematic of the IC demodulator.

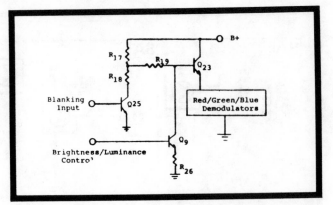

Fig. 2-13. Schematic of the blanker and brightness-luminance control stages.

from the power supply, cutting off this transistor, and the second color IF bias drops to zero, cutting it off as well.

However, with incoming color, and with the Instamatic button pushed to M (Fig. 2-16), the color killer is conducting current, and its positive output biases on Q4 in series with the emitter of Q3. Previously, Q3 had been securely shut down not only because the base was grounded through M, but also due to the absence of a return path to ground through the emitter. With the Instamatic switch in A (automatic), ground if lifted from voltage divider R3 and R9—Eo equals Ein X R2 divided by R1 + R2; Eo equals 20 X 15K divided by 47K + 15K or 300KV divided by 62K which is 4.84 volts. So Q3 is biased 4.84 volts positive, and turns on hard. Since Q3 and Q4 are in series coincidence, they can now be termed a digital AND gate, and when they conduct, the collectors are virtually at ground except for the internal resistance of each transistor.

With the coincidence AND gate in conduction, the top end of C4 is now shunted to ground, placing it in parallel with C41, which is already connected between inductor L8 and ground. By simple

geometric circuit analysis (Fig. 2-18) and the standard reactive equations, C41 (shown as 35 pf) has a reactance of 3790 ohms C4 a reactance of 1378 ohms, and parallel reactance (Xt) of the two amounts to some 1,000 ohms. Of course, you're not dealing with a simple capacitive-reactance since there is a series tuned circuit consisting of L7, C3, and L8, also shunted by C40 on the other end. So the addition of twice as much capacitance to a complex impedance automatically changes the phase angle and the lessened reactance to ground further loads the input to the blue demodulator. In addition, since the emitter current of the red video output is already set by R89 after being divided by R90 and R91, R4 shunted to ground through the 2-transistor AND gate now makes R90 not 330 ohms as before, but 330 ohms in parallel with 12,000 ohms or 292 ohms as the lower portion of the divider to ground. This drops the positive emitter voltage and causes additional

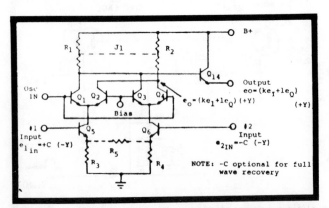

Fig. 2-12. Schematic of a single demodulator stage.

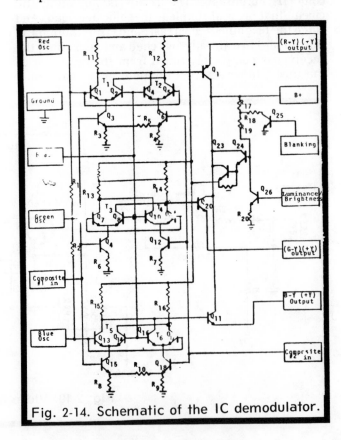

Fig. 2-14. Schematic of the IC demodulator.

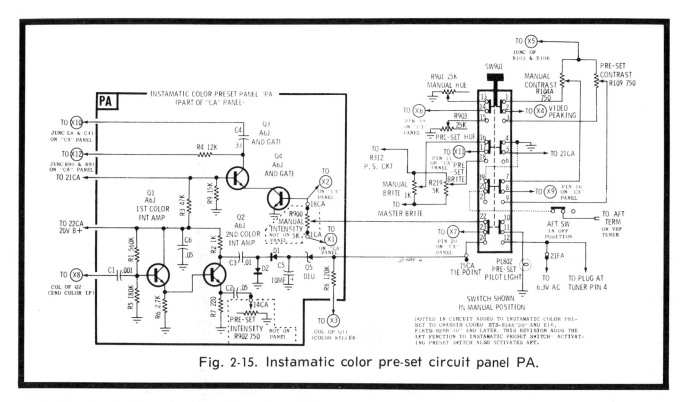

Fig. 2-15. Instamatic color pre-set circuit panel PA.

conduction of red video output Q17. The net result is that reds in the picture increase slightly and the demodulation angle is shifted, permitting broader fleshtone coverage of all televised scenes.

When color is received, the first and second internal color amplifiers continue to supply negative DC voltage (Fig. 2-17) to the collector of the color killer through Terminal 15, and the same potential through R58 (2.7K) to the base of second color IF Q2. This is simply an AC voltage-to-DC control 2-stage amplifier with bypass and resistive preset intensity in the emitter of the second internal amplifier that will control stage degeneration. A slight forward DC bias comes from the 20-volt supply through divider resistors R1 and R5 at the base of the first internal amplifier and is further reinforced by positive chroma from the collector of Q2. In the output of internal amplifier no. 2, the positive chroma alternations are shunted to ground

through D2 and negative alternations are passed by D1, filtered and supplied as a negative control source through D1U, zener diode Q5. With more incoming color, the output voltage through the internal amplifiers is increased and more back bias is applied to the second color IF with, of course, SW901 in the Instamatic position. When SW901 is moved to manual, the color killer output goes through the top portion of manual intensity potentiometer R900 and then to the second color amplifier to regulate its conduction. Meanwhile, the base of the first color IF (not shown) is always under control of the ACC amplifier through another DC-type bias arrangement. Therefore, with Instamatic, the color intermediate amplifiers are DC regulated for color sync and color level amplitudes, insuring complete automatic control of most video electronics except color phase, which is broad enough to cover many circumstances.

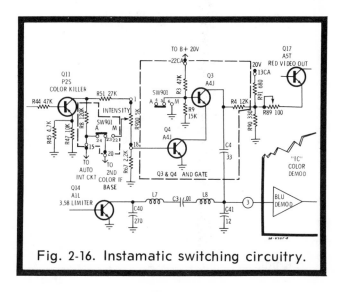

Fig. 2-16. Instamatic switching circuitry.

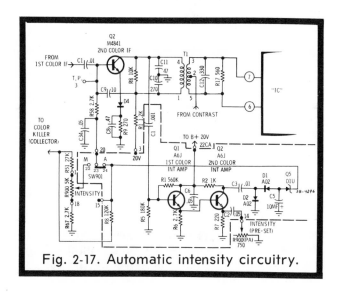

Fig. 2-17. Automatic intensity circuitry.

33

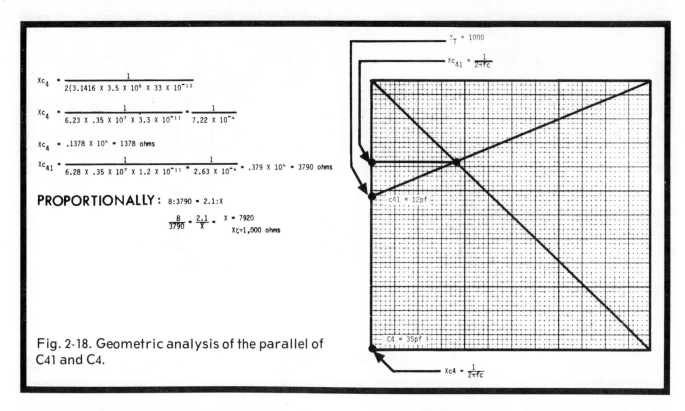

$$Xc_4 = \frac{1}{2(3.1416 \times 3.5 \times 10^6 \times 33 \times 10^{-12}}$$

$$Xc_4 = \frac{1}{6.23 \times .35 \times 10^7 \times 3.3 \times 10^{-11}} = \frac{1}{7.22 \times 10^{-4}}$$

$$Xc_4 = .1378 \times 10^4 = 1378 \text{ ohms}$$

$$Xc_{41} = \frac{1}{6.28 \times .35 \times 10^7 \times 1.2 \times 10^{-11}} = \frac{1}{2.63 \times 10^{-4}} = .379 \times 10^4 = 3790 \text{ ohms}$$

PROPORTIONALLY: $8 : 3790 = 2.1 : X$

$$\frac{8}{3790} = \frac{2.1}{X} = \quad X = 7920$$
$$X_\zeta = 1,000 \text{ ohms}$$

Fig. 2-18. Geometric analysis of the parallel of C41 and C4.

VIDEO OUTPUT STAGES, CTV5 & CTV6

The three video output stages (Fig. 2-19) are identical: linear amplifiers driven by a low-impedance signal from an emitter follower in the IC. In the emitter circuit of each amplifier a drive control sets the current through the transistor and thereby controls the collector voltage, the voltage applied to the CRT. Each drive control is connected to a voltage source by a divider between ground and the regulated 20-volt supply. The collectors are supplied a relatively constant voltage ($+200$) through 10K load resistors.

The simple regulator circuit in Fig. 2-19 provides a low-impedance, relatively constant voltage to the output transistors. The regulator transistor is protected from CRT arcing by diode D3. Should the regulator transistor short, the 3.3K resistor in the collector will still drop some of the B+ to the video outputs.

CRT CIRCUIT

The CTV5 uses a solid-state HV rectifier and provides up to 27KV DC to the CRT HV anode (Fig. 2-20). A resistor connected to the HV output provides a focus voltage of around 5KV that can be varied by the focus control. The G2 control, tied to the boost source, supplies a variable voltage to each G2 grid. The G1s are connected together and a tap provides either $+38$ or $+75$ volts. The video output transistors are coupled directly to the CRT cathodes.

The CTV6 CRT circuit is similar but requires a lower high voltage. The HV rectifier is a 3BW2 tube. This low-drive CRT uses a low focus voltage. Taps are provided as a means of selecting the voltage that provides optimum focus.

COLOR IF AMPLIFIER CTV5 & CTV6

Composite color (sidebands and sync) is applied to the first color IF (Fig. 2-21) from the color takeoff coil. The gain of the first IF varies with ACC circuit action as the color sync level increases or decreases. The output supplies both the color sync amplifier and second color IF. The gain of the second color IF is determined by the setting of the intensity control. Forward bias applied through the intensity control from the killer stage turns on the second IF. Through T1 two identical but 180-degree out-of-phase signals are applied to the IC demodulators. Composite video is applied to the T1 secondary centertap from the contrast control.

D1 in the first IF collector limits the amplitude variation of the signal caused by the flywheel action of L1. A diode with a forward voltage applied acts like a resistor in which the resistance varies inversely with the voltage across or through it.

When the intensity control is set at maximum, the second color IF conducts harder because of the increased forward bias. The increased current through Q2 also flows through D4, reducing the resistance proportionally. As a result, Q2 operates with maximum gain. As the forward bias is reduced (by reducing the intensity control setting), the resistance of the diode increases due to the decreasing current through it. Now, with the emitter no longer near AC ground, degeneration occurs across the diode and reduces the gain. Gain reduction is a function of the decreased forward bias plus the degeneration.

PULSE FORMER, LIMITER & COLOR SYNC AMPLIFIER, CTV5 & CTV6

The pulse former transistor (Q1 in Fig. 2-22) exhibits a low impedance because of its high for-

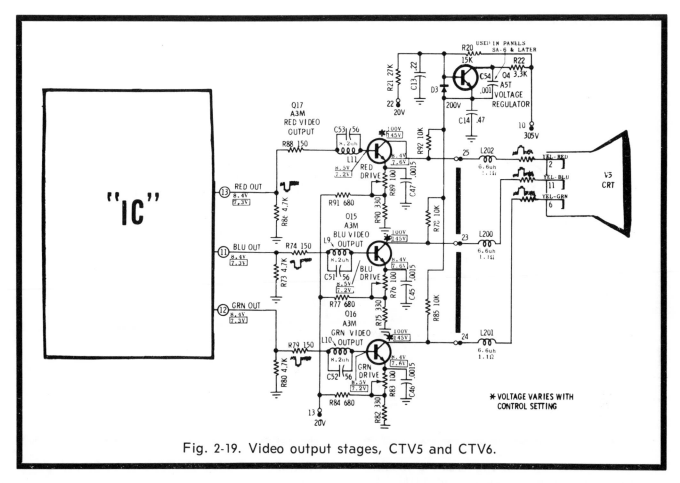

Fig. 2-19. Video output stages, CTV5 and CTV6.

ward bias. A pulse from the flyback is applied (through a 100K resistor) to the collector, but most of this pulse is dissipated across the resistor due to the shunting effect of the transistor. When a station signal is received, positive sync pulses are applied to the emitter of the pulse former transistor. This reduces conduction and forms a positive pulse at the collector. This pulse (at the 15,750-Hz rate) is direct coupled to the base of the pulse limiter and appears

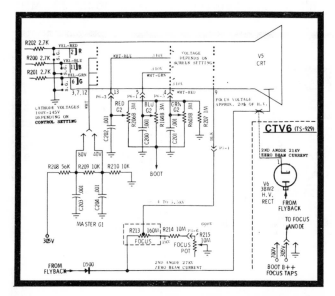

Fig. 2-20. CRT circuit for CTV5 receivers. The CTV6 circuit is the same, except for the differences shown within the dashed line.

inverted at the collector. This negative pulse is coupled through a series connected LC circuit (to introduce some delay) to the emitter of the color sync amplifier.

Because of the delay, the pulse formed from the sync is applied to the emitter, turning the transistor on coincident with the color sync on the base. The color sync amplifier will conduct only when turned on by the gating pulse and will have a 3.58-MHz output only if color sync is present on the base at the same time. The color sync signal appears at the output of the amplifier and is coupled to the base of the crystal driver transistor. The 27-pf capacitor passes 3.58-MHz readily but has a high reactance to the 15,750-Hz component.

CRYSTAL DRIVER, AMPLIFIER & 3.58-MHz OSCILLATOR

The crystal driver (Q7) receives the 3.58-MHz color sync signal from the gated color sync amplifier (Fig. 2-23). Q7 is an emitter follower which drives the crystal to supply a CW signal to the following stage. The 4.7-pf capacitor is approximately equal to the crystal capacity and, therefore, serves to neutralize the circuit. R36 reduces the circuit Q to increase the bandwidth. The color sync burst, and the resulting output from the ringing crystal, is applied to the crystal amplifier, which is made degenerative (and DC stabilized) by R40.

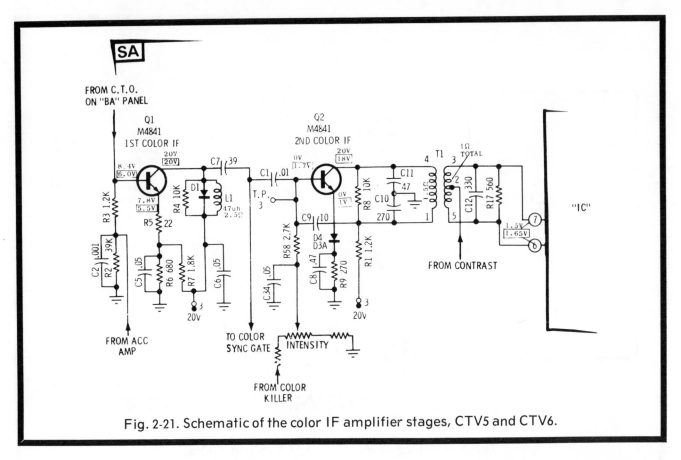

Fig. 2-21. Schematic of the color IF amplifier stages, CTV5 and CTV6.

The oscillator (Q10 in Fig. 2-23), a modified Colpitts, is free-running around 3.58 MHz. It is locked in sync with the incoming burst, applied through C24. The oscillator output is taken at the emitter through a low-impedance isolation network and applied to the phase splitter. Some phase correction is introduced by the 5.6-microhenry choke. The crystal output amplifier also feeds a 3.58-MHz signal to the ACC rectifier to develop a positive DC voltage to turn on the ACC amplifier. The only adjustment in these stages is the 3.58-MHz oscillator coil.

PHASE SPLITTER-SHIFTER, CTV5 & CTV6

The CW signal from the oscillator is coupled to the phase splitter which produces two output voltages 180 degrees apart. The CW signals at the emitter and collector are coupled to the emitter and collector of the phase shifter (Fig. 2-24). The phase shifter bias is determined by the setting of the hue control. With the arm of the control toward ground, no forward bias is applied so the output from the collector of Q12 is coupled directly to the 3.58-MHz

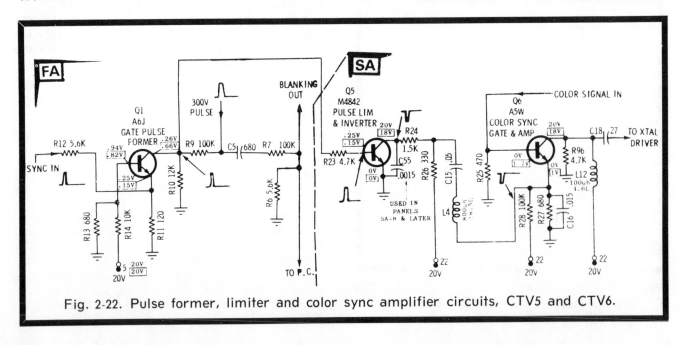

Fig. 2-22. Pulse former, limiter and color sync amplifier circuits, CTV5 and CTV6.

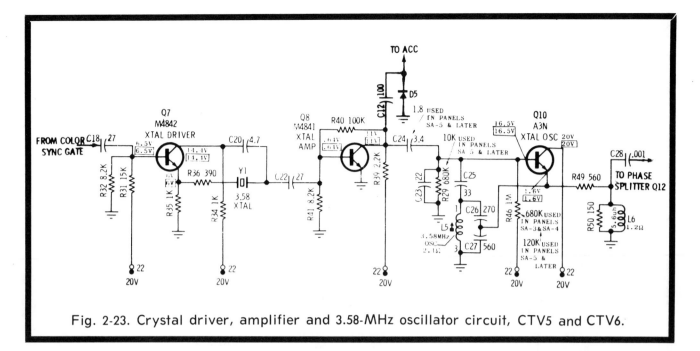

Fig. 2-23. Crystal driver, amplifier and 3.58-MHz oscillator circuit, CTV5 and CTV6.

limiter, which is relatively unaffected by the phase shifter in its cutoff condition.

Maximum forward bias is applied to the phase shifter with the hue control in the opposite extreme position. This makes the shifter transistor impedance very low, providing a path for the 3.58-MHz CW signal (on its emitter) through the transistor to the limiter, while shunting the high-impedance signal from the collector of the phase splitter.

With the hue control set between these two extremes, a portion of the out-of-phase signals from the collector and emitter of the phase splitter combine to produce a resultant output. The hue control is decoupled, allowing only DC on the control and base of the shifter. The hue control is continuously variable over a range of 150 degrees.

3.58-MHz LIMITER, CTV5 & CTV6

The purpose of the limiter is to supply a low-level (300 mv) 3.58-MHz CW signal to the IC (at the correct phase) for demodulation of the three color signals. It operates far into cutoff during most of the cycle, due to the charge on the .001 capacitor from Q13 (Fig. 2-25), and it draws very little average current (approximately 0.5 ma). C36 provides neutralization. Reference level changes that can occur as a result of hue control adjustment will cause a shift in the bias on the limiter to maintain a constant output. Since the IC CW signal requirements are small, the output is taken from the low end of the tuned circuit. The tuned circuit in the output allows recovery of the complete 3.58-MHz sine wave and also provides a method of adjusting the hue control range.

ACC AMPLIFIER & COLOR KILLER, CTV5 & CTV6

The ACC amplifier is the first stage of a solid-state switch (Fig. 2-26). In the absence of a color

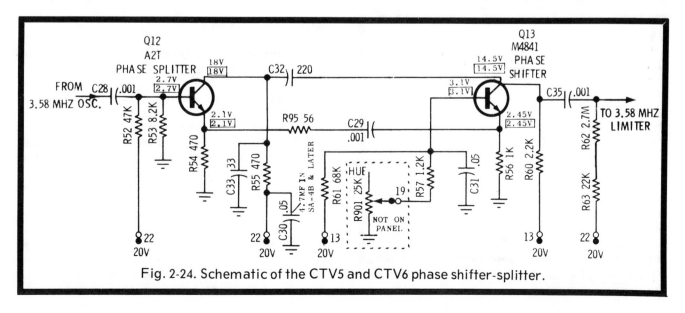

Fig. 2-24. Schematic of the CTV5 and CTV6 phase shifter-splitter.

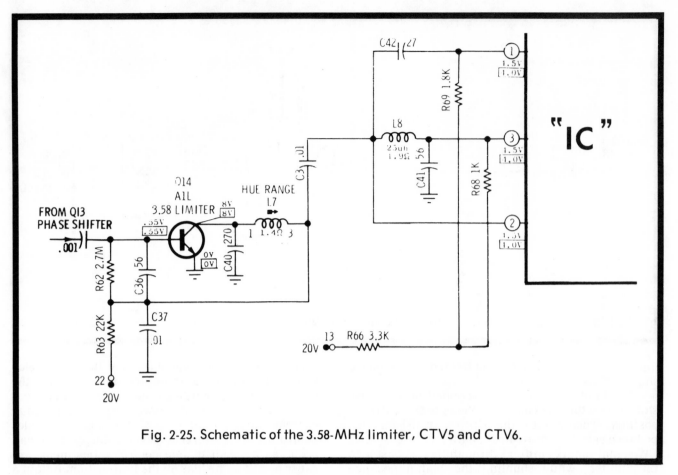

Fig. 2-25. Schematic of the 3.58-MHz limiter, CTV5 and CTV6.

signal, neither stage conducts. When a color signal is received, a 3.58-MHz signal from the crystal output amplifier is present at the ACC rectifier (D5). The resulting positive voltage turns on the ACC amplifier and the collector voltage drops. Direct coupled to the next stage, this negative-going voltage turns on the killer transistor (PNP) and the collector becomes positive. To improve the action of this circuit, it is made regenerative by a resistor connected from the collector of the killer back to the base of the ACC amplifier. The collector of the killer is returned to ground through the intensity control. Positive voltage developed at the collector biases on

the second color IF amplifier. The gain in the second color IF is determined by the setting of the intensity control.

Bias for the first color IF is set up by voltage dividers to provide maximum gain, but it is automatically reduced, depending on the color sync amplitude. The DC voltage on the collector of the ACC amplifier is proportional to the amplitude of the 3.58-MHz CW signal applied to the ACC rectifier. As this amplitude increases, the collector voltage on the ACC amplifier decreases, forward biases the ACC delay diode and reduces the forward bias of the first color IF transistor, which reduces its gain.

AGC, CTV5 & CTV6

Forward AGC is applied to the first IF transistor for gain reduction and delayed AGC is provided for the RF amplifier. Composite video (negative-going sync) is applied to the base of AGC gate Q3 in Fig. 2-27. The stage is reverse biased until the negative-going sync turns it on.

A negative pulse from a winding on the flyback transformer is applied to the collector of the gate through a diode. The stage conducts when the sync and flyback pulses arrive at the same time, and the degree of conduction is determined by the amplitude of the signal on the base and the setting of the AGC control.

When the gate conducts, C2 charges with the polarity shown. During scan time the gate is not conducting and the charge on C2 is drained off by R10. This forward biases Q2 and the current through

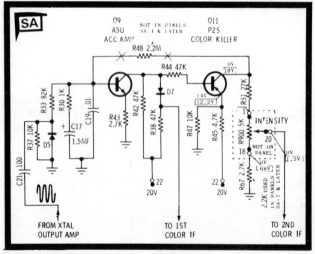

Fig. 2-26. CTV5 and CTV6 ACC amplifier and color killer circuits.

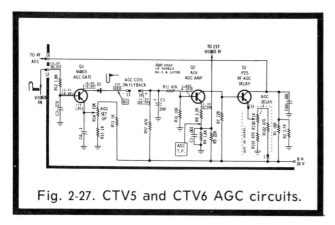

Fig. 2-27. CTV5 and CTV6 AGC circuits.

the emitter resistor makes the emitter positive. The resulting voltage is applied to the first video IF and varies as a function of incoming signal strength.

As current increases in the AGC amplifier, the collector voltage drops. The collector is direct coupled to the base of the RF AGC delay transistor, which is normally cut off. When a signal of sufficient strength is received, the base voltage drops far enough to turn the stage on, making the collector more positive. This positive voltage, applied to the tuner, reduces the gain of the RF amplifier. A control is provided in the emitter of the RF AGC delay transistor to set the AGC point.

AUDIO IF CTV5 & CTV6

Because it is desirable to keep 4.5-MHz out of the video, the 41.25-MHz audio carrier is taken out by traps ahead of the video detector. The 45.75-MHz video and 41.25-MHz audio carriers are picked off the third IF through a small capacitor. The audio 4.5-MHz IF is produced by mixing the video carrier with the frequency modulated audio carrier in a diode. The output of the diode mixer is tuned to 4.5-MHz by a tunable coil, tapped to minimize loading and provide coupling to the IF amplifier in the IC. The mixer diode is slightly forward biased through the IC for improved sensitivity and more linear operation (Fig. 2-28). The recovered audio at the output of the detector is available at Terminal 2 and goes to the tone and volume controls. The audio is

returned to the IC at Terminal 9 for low-level amplification. The audio from the IC is direct coupled from an emitter follower stage at Terminal 10 for impedance matching into the first stage of the audio amplifier.

AUDIO AMPLIFIER, CTV5 & CTV6

DC coupling is used between all transistors (Fig. 2-29). Negative feedback, provided by R50, minimizes distortion at high volume settings and provides DC stability.

HORIZONTAL OSCILLATOR & DRIVER, CTV5 & CTV6

The horizontal oscillator (Fig. 2-30) is a modified Hartley. AFC voltage locks the oscillator with the incoming sync in the conventional manner. The oscillator output is direct coupled to the driver grid. After shaping and amplification in the driver, the signal is applied to the output grid.

HORIZONTAL OUTPUT, CTV5 & CTV6

The horizontal output tube (Fig. 2-31) uses grid leak bias, developed as a result of the high amplitude drive signal. The grid curve on the positive excursions builds a charge on C27 and leaves an average negative 60 to 90 on the grid. A supplementary bias is developed by a feedback circuit (C501, R506, VDR R502 and R503) from the horizontal output transformer. A pulse from the flyback is coupled to R502 by C501. This increased voltage (positive-going pulse) causes the VDR resistance to decrease while the pulse is present, placing a charge on C501. The net charge on the capacitor increases the bias on the output tube through R501. Since the amplitude of the pulse is proportional to the high voltage, the circuit will react to provide regulation for sweep and high voltage. R503 provides a means of HV adjustment. R503 is adjusted at the factory to 27KV maximum (21KV on CTV6) at a 122v AC line. Windings on the horizontal output transformer provide pulses for the AGC gate, horizontal AFC, pulse former and convergence circuits.

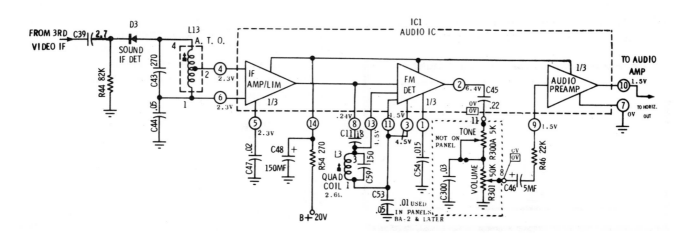

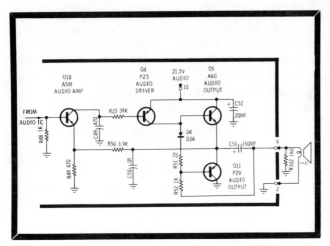

Fig. 2-29. Schematic of the CTV5 and CTV6 audio amplifier circuits.

VERTICAL OSCILLATOR & OUTPUT, CTV5 & CTV6

The vertical oscillator and output stages are similar to those used in previous Motorola tube-type color receivers (Fig. 2-32). V3B is a free-running multivibrator. The signal from the sync separator is applied to the integrator and coupled to the oscillator to lock the oscillator on frequency. The vertical output tube amplifies the sawtooth voltage applied and supplies current to drive the yoke, through the output transformer. Voltage occuring at vertical retrace time is coupled back to the grid of the oscillator to sustain oscillation. B+ for the oscillator is supplied from the boost circuit and regulated by a voltage-dependent resistor (VDR). The vertical size and linearity controls are connected between the regulated voltage and the negative 35-volt supply.

The vertical output stage operates from the 260-volt supply. Line voltage variations can cause this voltage to vary with a corresponding change in vertical size. However, the negative 35 volts also changes with a change in line voltage. This voltage applied to the vertical oscillator plate circuit corrects for size change that would result from B+ changes. The cathode of the vertical output tube is bypassed by a 50-mfd capacitor on the convergence panel. The vertical output transformer has separate windings for the yoke and convergence circuitry.

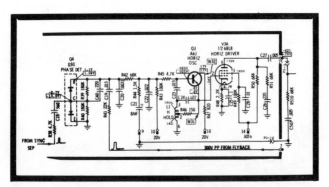

Fig. 2-30. Horizontal oscillator and driver stages, CTV5 and CTV6.

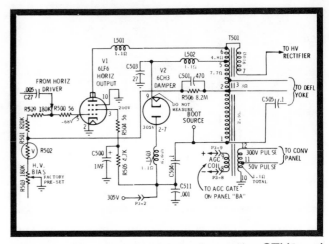

Fig. 2-31. Horizontal output schematic, CTV5 and CTV6.

CTV5 POWER SUPPLY

The power supply (Fig. 2-33) has two basic B+ supplies. A low-voltage winding of the power transformer supplies 27 volts AC to a single silicon rectifier on the ZA panel and associated circuitry to provide 21.5 volts DC for the audio B+. The DC developed at the cathode of the rectifier is also applied to a low-voltage regulator which provides a regulated output that is adjustable to 20 volts for normal operation.

A higher voltage secondary winding is used in a full-wave voltage-doubler circuit to supply 260 volts for the vertical output. High B+ for the remainder of the tube-type sweep circuits is taken from the same doubler circuit, but uses a choke in the filter network. The video output transistors also receive B+ from the 305-volt supply.

Two 6.3-volt windings are used on the power transformer: one winding supplies filament voltage to the tubes and pilot lamps, and the second is isolated from the chassis to supply 6.3 volts AC to the CRT filament.

AC voltage for automatic degaussing is taken from the high-voltage secondary winding. The degaussing coils are in series with a resistor which has a positive temperature coefficient. When cold, the resistance is quite low. Circuit current causes the resistance to rise rapidly. AC flows in the coils each time the set is turned on, but quickly diminished as the resistor value increases. The power supply is protected by a conventional type of circuitbreaker (3.5 amp) in the primary of the power transformer.

CTV6 POWER SUPPLY

The CTV6 chassis is operated from the power line and uses a polarized line plug. A doubler circuit (Fig. 2-34) furnishes 305 and 260v B+ sources. R802 provides surge protection.

Filaments are supplied by a 6.3v transformer, with a separate winding for the CRT. A tap on the transformer primary winding develops 27 volts for the low B+ supply and the 35-volt negative source.

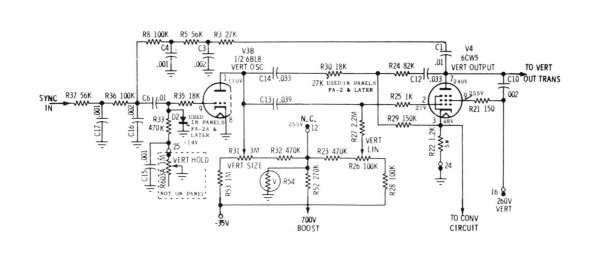

Fig. 2-32. Vertical oscillator and output circuits, CTV5 and CTV6.

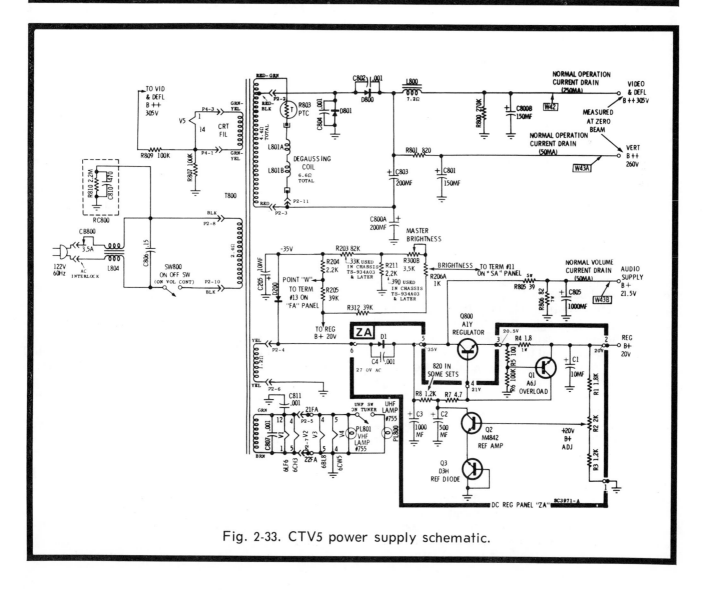

Fig. 2-33. CTV5 power supply schematic.

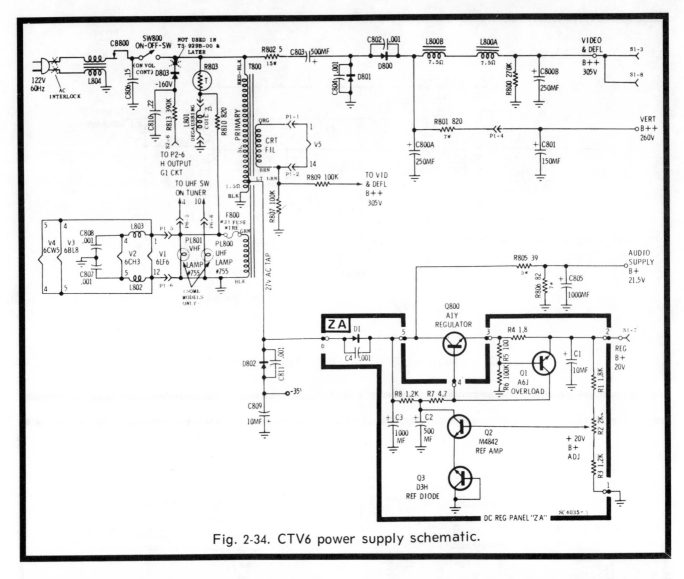

Fig. 2-34. CTV6 power supply schematic.

The supply is protected by a 3.5-amp breaker in the AC line. The filament circuit is protected by a no. 31 wire fuse.

REGULATED 20-VOLT DC SUPPLY, CTV5 & CTV6

The transformer winding provides 27 volts AC to the silicon diode on the ZA panel (Figs. 2-33 and 2-34). The rectified output voltage is filtered by a 1000-mfd capacitor and applied to the collector of off-panel regulator transistor Q800. A portion of the output voltage, applied to the base, determines Q2's conduction. Any change in output load or supply input voltage does not appear at the output voltage, for the regulator acts to oppose such a change. For instance, if the output voltage starts to drop, the Q2 base voltage drops, reduces the conduction of Q2 and causes its collector voltage to rise. This increases conduction of Q800 (less resistance) and brings the output voltage back up. Thus, any increase or decrease is automatically corrected. Control R2 provides a means of adjusting the output to 20 volts. Improved regulation and temperature stability are achieved using reference diode D2. Current limiting is provided by Q1, which is connected across the emitter and base of Q800.

Q1 is normally not conducting with the normal 250 ma through R4 (1.8 ohms), but as current increases, more drop across R4 brings Q1 into conduction. At 450 ma, Q1 is conducting hard and limits the maximum current to protect the power supply components.

INSTANT-ACTION, TS-929-934

The STS-934 receiver incorporates "Instant-Action," which eliminates warmup time. A relay has been added to the receiver, and an additional on-off switch is ganged to the original switch (Fig. 2-35).

With the receiver off, a 50-ohm resistor across the on-off switch supplies reduced power to the transformer primary. The tubes in the receiver (CRT and sweep circuit) are preheated. When the on-off switch is closed, the 50-ohm resistor is shorted out and full power is applied to the transformer primary. All tubes now operate with full power. The additional on-off switch closes and AC is applied to diode D1 and regulator panel ZA. DC is applied to the relay coil, closing two contacts. One contact completes the circuit to the degausser and voltage-doubler circuits. The other relay contact closes and

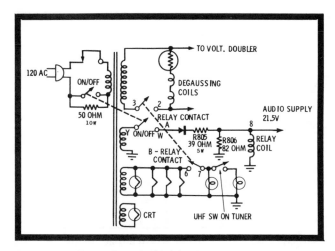

Fig. 2-35. Schematic of the Instant-Action circuit, CTV5 and CTV6.

applies AC to the channel indicator lamps for VHF and to the UHF indicator lamp when in the UHF position.

SERVICE & SETUP

For the CTV5 chassis, purity, convergence and CRT replacement instructions appear in Fig. 2-36. The dynamic convergence procedure is presented in Fig. 2-37. Fig. 2-38 explains the service adjustments, and Figs. 2-39, 2-40, 2-41 and 2-42 show various chassis views. In Fig. 2-43, you'll find panel removal and general service information, and Fig. 2-44 lists panel replacement and trouble symptoms.

For CTV6 receivers, purity, convergence and CRT replacement instructions are presented in Fig. 2-45. The dynamic convergence procedure appears in Fig. 2-46 and service adjustments in Fig. 2-47. Figs. 2-48 through 2-52 show various chassis views. A panel replacement troubleshooting chart is presented in Fig. 2-53.

Printed circuit board drawings in Figs. 2-54 through 2-59 show panel connections and component locations.

QUASAR II PROBLEMS

In the CTV5 chassis, when replacing **diode D200 and capacitor C205**, use 48S191A05 and 23S10255A60, a 100-mfd capacitor at 63v DC or better. In the CTV6 chassis this diode is D802 and the capacitor is C809; replacements are the same as given above.

Pin no. 1 on the ZA panel is the ground return and it must have secure contact or hum bars will develop in the picture. Should use a jumper with a grounding clip, available as a field kit no. 39P65175A63.

Hum in picture resembling filter hum bars can be eliminated by shunting the power supply diodes with 0.001-mfd disc ceramic capacitors to dissipate AC transients. Keep the leads short as possible. To accurately identify any power supply irregularities, use your oscilloscope and a 10 X LC probe to follow the ripple and line voltage to their points of difficulty. This trouble applies to both the CTV5 and 6.

In both chassis, diode D802 and capacitor C809 in the negative supply sometimes give trouble. Apparently what's needed is a negatively rectifying half-wave diode with a better peak inverse voltage rating and more current handling ability. Diode failure could be due to leakage in C809, which, by the way, can open and cause AC ripple on the negative supply. If it shorts and draws all sorts of current through D802, the circuitbreaker will be tripped. Usually, though, the capacitor simply opens or D802 opens.

On SA-8 panels and later, color intensity variations on different stations can be improved by adding a 560-pf capacitor from the Q5 collector to ground, since this changes the time constant of the gating pulse and improves AGC action.

When 0.5-amp slow-blow fuse F500 pops for some unknown reason, check the CRT beam current and adjust for 1.3 ma. In the field, turn both contrast and brightness controls to minimum, check the boost voltage with meter, then turn the brightness control to maximum. Adjust the master brightness for a boost voltage 50 volts **lower** than the initial reading with both contrast and brightness controls turned down. This corresponds to 1.3 ma of beam current.

Dark vertical bars may sometimes be noticeable in the CTV6 on the left of the CRT or on a blank raster. Replace 27-pf C503 at the damper cathode with a 68-pf 5 kilovolt disc capacitor. Should the bars still appear, shunt a 250-pf 5 KV capacitor between Terminals 1 and 2 on the flyback.

A brightness problem—low or none, check D200.

Hum bar in the picture—check C205, or look for a poor ground on the ZA panel.

Raster won't blank with the brightness control retarded. The G2s may be set too high, the wrong voltage tap on the CRT grid may be in use, or the spark gap on the CRT socket may be shorted. Also, the kine drive may be set too high.

Vertical sweep insufficient—check R508.

Horizontal output cathode fuse blown on TS-929. The master brightness control may be set too high, causing high beam current, which overworks the horizontal output.

CTV5 CHASSIS CODING CHANGES

TS-934A16: Audio jacks addition and electrolytic capacitor change. Two phono plugs, part number 9C67349A07, added to the rear of the chassis drawer to permit playing TV sound through an external amplifier. Capacitor C205 in the -35 volt brightness supply changed from 23C62914A14, 10 mfd, -0 + 250 percent 50v to 23C65808A37, 10 mfd, 10 + 100 percent 50v to improve reliability.

TS-934A18: Improving regulator panel ZA grounding to prevent hum bars in picture. Two ground clips (part number 42A70106A01) were added over the two nylon mounting grommets located on corners of the ZA panel opposite to the contact pins. Plating surfaces at these two corners must be cleaned and tinned to assure good contact with the clips.

STATIC CONVERGENCE

USE CROSSHATCH OR DOT PATTERN

1. Adjust red and green static magnets to converge pattern at center of screen.

2. Adjust blue static and blue lateral magnets to superimpose blue on red/green pattern at center of screen.

Some interaction may occur. Repeat steps 1 and 2 as necessary.

PURITY ADJUSTMENT

NOTE: CHECK STATIC CONVERGENCE AND CONVERGE IF NECESSARY.

1. Reduce blue and green G2 controls to minimum.

2. Loosen front and rear yoke mounting clamp.

3. Slide yoke forward or backward to obtain a red ball on CRT.

4. Spread ears of purity ring only as much as necessary, and rotate complete assembly to center red ball on CRT screen.

5. Position yoke for most uniform red field. Adjust purity ring as necessary to obtain red field.

6. Turn down red G2 control and alternately turn up green G2 and blue G2 controls. Observe individual fields, adjust purity ring as necessary to obtain pure individual fields. In some cases it may be necessary to re-adjust yoke to optimize purity of individual red, blue and green fields.

RED BALL

CHECK STATIC CONVERGENCE: reset if necessary. Recheck individual fields. Secure deflection yoke clamps. Proceed with background adjustment.

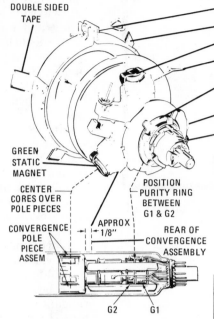

DOUBLE SIDED TAPE

FRONT YOKE MTG CLAMP

MTG EARS

BLUE STATIC MAGNET

REAR YOKE MTG CLAMP

PURITY RING

BLUE LATERAL MAGNET

RED STATIC MAGNET

GREEN STATIC MAGNET

CENTER CORES OVER POLE PIECES

CONVERGENCE POLE PIECE ASSEM

POSITION PURITY RING BETWEEN G1 & G2

APPROX 1/8"

REAR OF CONVERGENCE ASSEMBLY

G2   G1

COLOR TUBE ASSEMBLY AND ADJUSTMENTS

ASSEMBLING CRT

DEFLECTION YOKE MOUNTING

The following procedure is essential to assure accurate lateral yoke movement necessary to obtain optimum purity.

1. Position yoke firmly against CRT bell and secure rear yoke clamp.

2. Position front yoke mounting clamp onto front of yoke. Snug tighten clamp onto yoke. Slide clamp onto CRT bell and form clamp ears to bell of CRT.

3. Loosen rear yoke clamp to allow yoke movement for purity adjustment.

4. Tighten front and rear yoke clamps after tilt and purity adjustment.

CONVERGENCE YOKE AND BLUE LATERAL ASSEMBLY MOUNTING

1. Install yoke with cores centered over convergence pole pieces as shown in detail, and secure.

2. Install blue lateral assembly with respect to G1 & G2 as shown in detail and secure.

CRT REMOVAL AND INSTALLATION

CRT is mounted at front of cabinet.

1. REMOVE SPEAKER PANEL to gain access to screws securing control escutcheon. Remove control escutcheon. See label on chassis or service information sheet packed with set.

2. REMOVE CRT BEZEL. Bezel is secured with screws at corners of bezel readily accessible from rear of set.

3. REMOVE 4 CRT MOUNTING BOLTS at corners of CRT at front of set. Remove bottom bolts first. Support CRT while removing top bolts.

4. REMOVING AND INSTALLING CRT MOUNTING STRAP AND BRACKETS

Note position of strap and mounting brackets on CRT before disassembly. Read all notes on detail, use notes applicable to CRT being replaced (21" or 25").

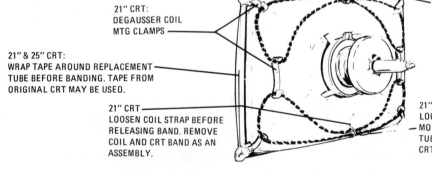

21" CRT:
DEGAUSSER COIL MTG CLAMPS

21" & 25" CRT:
WRAP TAPE AROUND REPLACEMENT TUBE BEFORE BANDING. TAPE FROM ORIGINAL CRT MAY BE USED.

21" CRT
LOOSEN COIL STRAP BEFORE RELEASING BAND. REMOVE COIL AND CRT BAND AS AN ASSEMBLY.

MTG BRACKET (4)

21" & 25" CRT:
MARK POSITION OF BRACKETS WITH RESPECT TO BAND OR TAPE TO BAND AND REMOVE AS AN ASSEMBLY.

21" & 25" CRT:
LOOSEN SCREW TO RELEASE BAND. MOUNT ASSEMBLY ON REPLACEMENT TUBE, POSITION BRACKETS AS ON ORIGINAL CRT AND TIGHTEN.

Fig. 2-36. CTV5 purity, convergence and CRT replacement instructions.

TS-934A20: To reduce brightness change with contrast. R106, located at the low end of the contrast control, changed from 4.7K to 5.6K 5 percent ½w.

TS-934A23: To reduce 3.58-MHz interference in the video response. Compensating coil L100 was changed from 220uh to 100uh (Part number 24D68002A28). C101 was changed from 220 pf to 100 pf 10 percent Z5F (Part number 21S180A98). Color-video panels coded SA-15 were incorporated in the chassis to correct the burst gate timing. This change consists of a value change in C55 from .0015 mfd to 560 pf 10 percent Z5F.

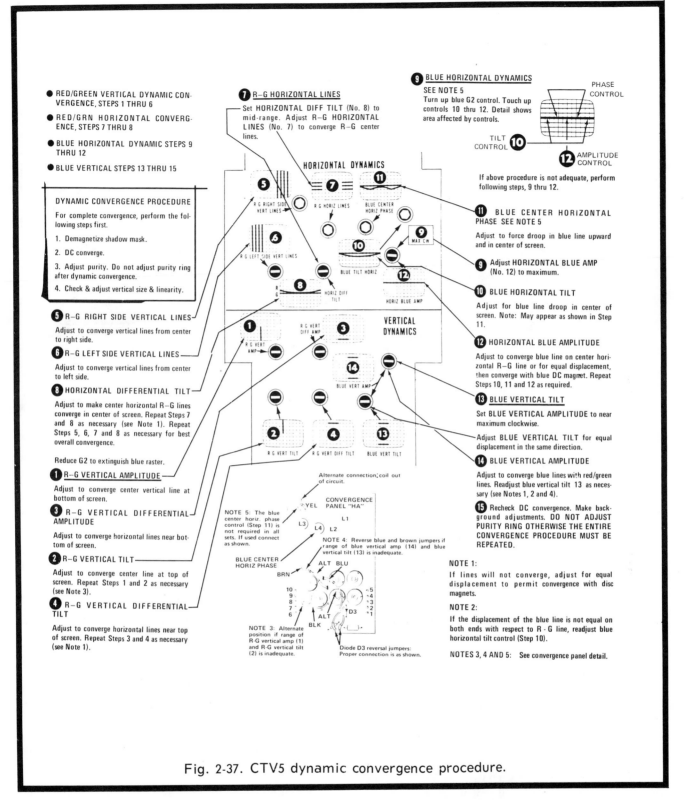

Fig. 2-37. CTV5 dynamic convergence procedure.

TS-934A24: Reliability change. Capacitor C805 1000 mfd, 25v, part number 23S10255A03, changed to a higher temperature unit of the same capacity, part number 23S10255A35.

TS-934A26: To reduce silicon bar. Capacitor C206, .001 mfd +80 -20 percent Z5U added across diode D200 in the -35v supply.

TS-934A-30: Revision to add AFT function to instamatic: Ground lead on the switch of the AFT on-off switch is removed from ground and wired through Contacts 19 and 20 of preset switch SW901. A jumper is added between Terminal 19 and grounded Terminal 4 to complete the return path to chassis. In the automatic position, Contacts 19 and 20 are open, activating the AFT circuit.

POWER SUPPLY, PS-934-2: To improve pincushion correction, the yoke leads to Terminals 5 and 6 should be reversed. P1-1 is connected to Terminal 5 and P1-5 to Terminal 6.

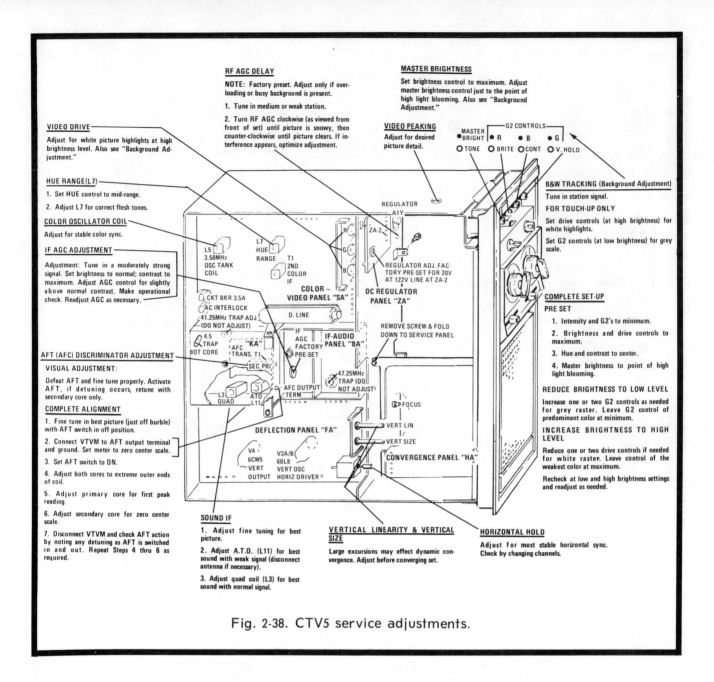

Fig. 2-38. CTV5 service adjustments.

PS-934-3: To center the vertical size pot adjustment, R508 is changed from 150K to 82K 10 percent 1w.

PS-934-4: Reliability change. R510 changed from 68K, 10 percent ½w to 150K, 10 percent ½w.

CTV6 S3 CHASSIS CODING CHANGES

TS-929B-07: Reliability change; C809 changed from 23C62914A14 (10 mfd, -0 + 250 percent, 50v) to 23C65808A37, (10 mfd, -10 + 100 percent, 50v).

TS-929B08: To eliminate AM rectification: C300, (470 pf, 20 percent, Z5F), added from the arm of the volume control to ground.

TS-929B-09: To eliminate yoke ringing: C507 (.001 mfd, 20 percent, Z5F), replaced by 1.0 mfd, 10 percent, 400v capacitor.

TS-929B-11: Improving regulator panel ZA grounding to prevent hum bars in picture: Two ground clips (Part No. 42A69179A01) are added over the two nylon mounting grommets located on the corners of the ZA panel opposite to the contact pins. Plating surface at these two corners must be cleaned and tinned to assure good contact with the clips.

TS-929B-12: To center the high-voltage range: R509, located on the PCC panel, changed from 39K (10 percent, 1w) to 27K (10 percent, 1w).

TS-929B-13: To center the high-voltage range: R506, changed from 6.8 meg to 8.2 meg, 10 percent, 1w.

TS-929B-17: Reliability change: C805 changed from Part No. 23S10255A03 to 23S10255A35. The new capacitor has an increased temperature rating.

TS-929B-20: Fuse sleeving change: Sleeving length changed from 1½ to ¾ inches to improve fuse heat dissipation.

TS-929B-21: To limit beam current: R218 (1200 ohms, 10 percent, ½w) added in series with the master brightness and brightness controls.

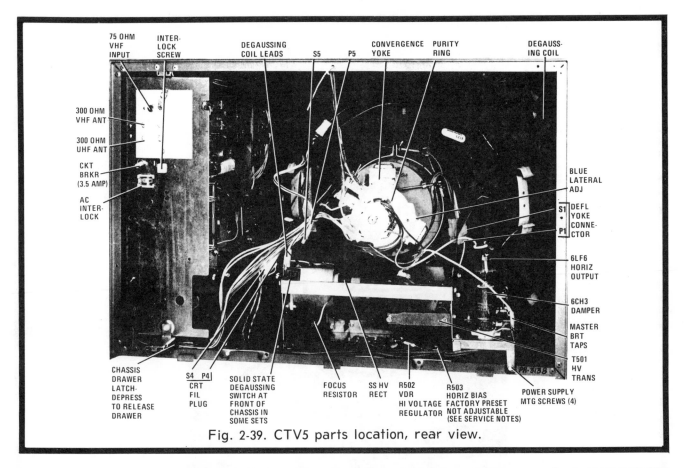

Fig. 2-39. CTV5 parts location, rear view.

Labels (Fig. 2-39):
- 75 OHM VHF INPUT
- INTER-LOCK SCREW
- DEGAUSSING COIL LEADS
- S5
- P5
- CONVERGENCE YOKE
- PURITY RING
- DEGAUSSING COIL
- 300 OHM VHF ANT
- 300 OHM UHF ANT
- CKT BRKR (3.5 AMP)
- AC INTERLOCK
- BLUE LATERAL ADJ
- S1 / P1 DEFL YOKE CONNECTOR
- 6LF6 HORIZ OUTPUT
- 6CH3 DAMPER
- MASTER BRT TAPS
- T501 HV TRANS
- CHASSIS DRAWER LATCH-DEPRESS TO RELEASE DRAWER
- S4 P4 CRT FIL PLUG
- SOLID STATE DEGAUSSING SWITCH AT FRONT OF CHASSIS IN SOME SETS
- FOCUS RESISTOR
- SS HV RECT
- R502 VDR HI VOLTAGE REGULATOR
- R503 HORIZ BIAS FACTORY PRESET NOT ADJUSTABLE (SEE SERVICE NOTES)
- POWER SUPPLY MTG SCREWS (4)

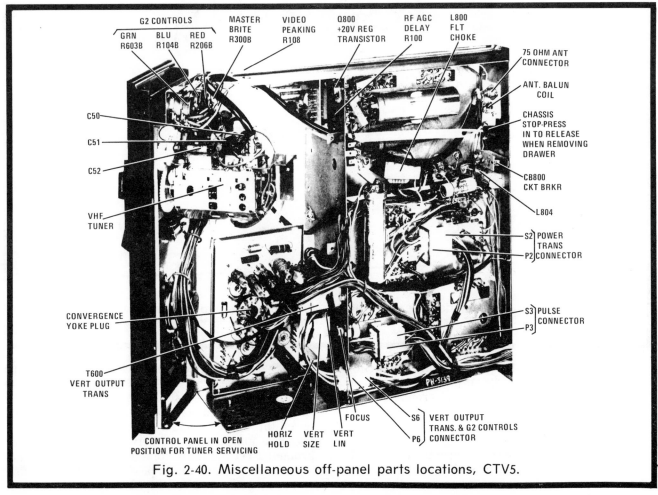

Fig. 2-40. Miscellaneous off-panel parts locations, CTV5.

Labels (Fig. 2-40):
- G2 CONTROLS: GRN R603B, BLU R104B, RED R206B
- MASTER BRITE R300B
- VIDEO PEAKING R108
- Q800 +20V REG TRANSISTOR
- RF AGC DELAY R100
- L800 FLT CHOKE
- 75 OHM ANT CONNECTOR
- ANT. BALUN COIL
- CHASSIS STOP-PRESS IN TO RELEASE WHEN REMOVING DRAWER
- C50
- C51
- C52
- CB800 CKT BRKR
- L804
- VHF TUNER
- S2 / P2 POWER TRANS CONNECTOR
- S3 / P3 PULSE CONNECTOR
- CONVERGENCE YOKE PLUG
- T600 VERT OUTPUT TRANS
- CONTROL PANEL IN OPEN POSITION FOR TUNER SERVICING
- HORIZ HOLD
- VERT SIZE
- VERT LIN
- FOCUS
- S6 / P6 VERT OUTPUT TRANS. & G2 CONTROLS CONNECTOR

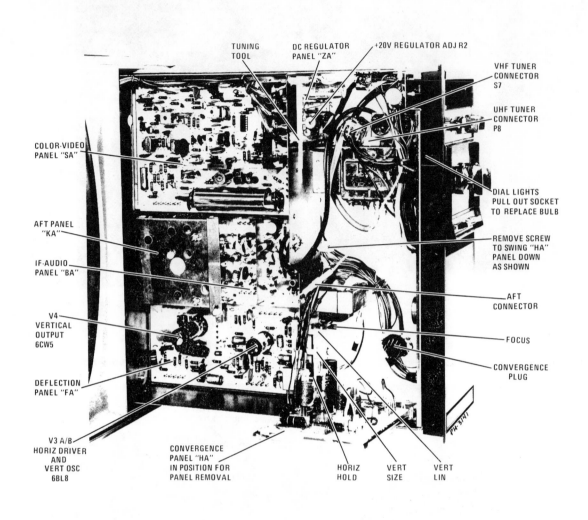

TUNING TOOL — DC REGULATOR PANEL "ZA" — +20V REGULATOR ADJ R2

VHF TUNER CONNECTOR S7

UHF TUNER CONNECTOR P8

COLOR-VIDEO PANEL "SA"

DIAL LIGHTS PULL OUT SOCKET TO REPLACE BULB

AFT PANEL "KA"

IF-AUDIO PANEL "BA"

REMOVE SCREW TO SWING "HA" PANEL DOWN AS SHOWN

AFT CONNECTOR

V4 VERTICAL OUTPUT 6CW5

FOCUS

CONVERGENCE PLUG

DEFLECTION PANEL "FA"

V3 A/B HORIZ DRIVER AND VERT OSC 6BL8

CONVERGENCE PANEL "HA" IN POSITION FOR PANEL REMOVAL

HORIZ HOLD

VERT SIZE

VERT LIN

Fig. 2-41. Panel identification and off-panel component locations, CTV5.

TS-929B-24: To reduce silicon bar: Capacitor C814, (.002 mfd, Part No. 21S115383) added between the collector of regulator Q800 and ground.

TS-929B-28: To reduce shading in the raster: RF choke L505, 6.6uh, added in series with +305-volt line to the horizontal output tube.

TS-929B-30; affects only chassis that feature Insta-Matic preset: Revision to add AFT function to Instamatic preset switch. The ground lead of the AFT off-on switch is removed from ground and wired through Contacts 19 and 20 of preset switch SW901. A jumper is added between Terminal 19 and grounded Terminal 4 to complete the return path to chassis. In the Instamatic position, Contacts 19 and 20 are open, activating the AFT circuit.

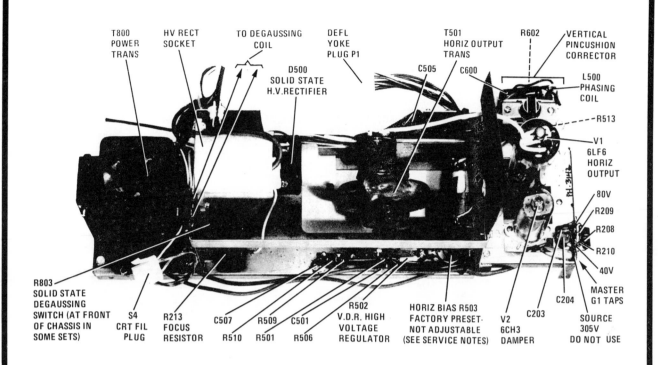

T800 POWER TRANS

HV RECT SOCKET

TO DEGAUSSING COIL

DEFL YOKE PLUG P1

D500 SOLID STATE H.V. RECTIFIER

T501 HORIZ OUTPUT TRANS

C505

C600

R602

VERTICAL PINCUSHION CORRECTOR

L500 PHASING COIL

R513

V1 6LF6 HORIZ OUTPUT

80V

R209

R208

R210

40V

MASTER G1 TAPS

SOURCE 305V DO NOT USE

R803 SOLID STATE DEGAUSSING SWITCH (AT FRONT OF CHASSIS IN SOME SETS)

S4 CRT FIL PLUG

R213 FOCUS RESISTOR

C507

R510

R509

R501

C501

R506

R502 V.D.R. HIGH VOLTAGE REGULATOR

HORIZ BIAS R503 FACTORY PRESET- NOT ADJUSTABLE (SEE SERVICE NOTES)

V2 6CH3 DAMPER

C203

C204

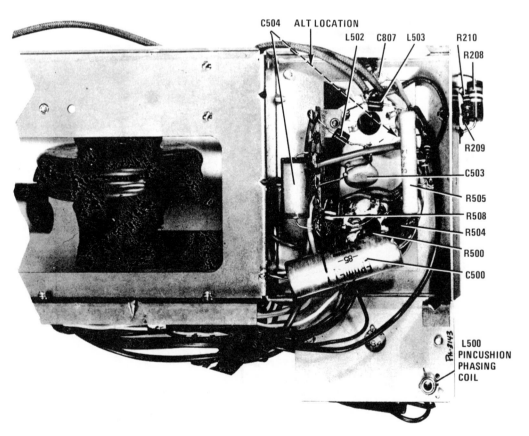

C504 ALT LOCATION

L502 C807 L503

R210

R208

R209

C503

R505

R508

R504

R500

C500

L500 PINCUSHION PHASING COIL

Fig. 2-42. Power supply component location (top) and horizontal output component locations, CTV5.

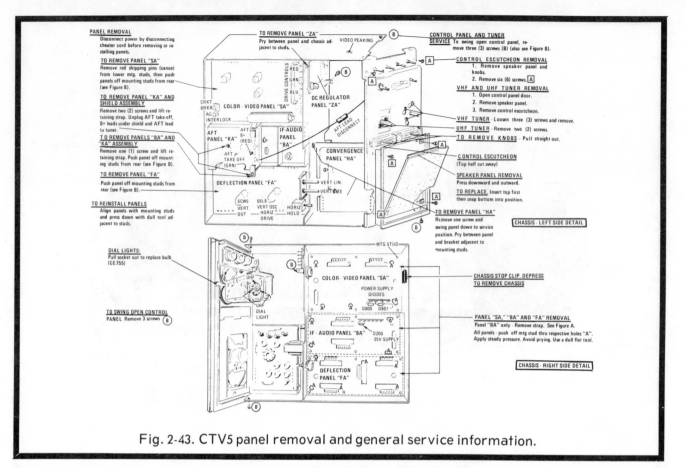

Fig. 2-43. CTV5 panel removal and general service information.

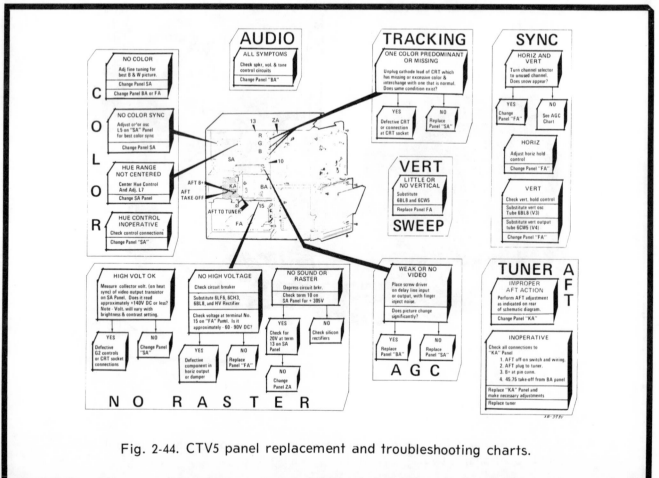

Fig. 2-44. CTV5 panel replacement and troubleshooting charts.

STATIC CONVERGENCE

USE CROSSHATCH OR DOT PATTERN

1. Adjust red and green static magnets to converge pattern at center of screen.

2. Adjust blue static and blue lateral magnets to superimpose blue on red/green pattern at center of screen.

Some interaction may occur. Repeat steps 1 and 2 as necessary.

PURITY ADJUSTMENT

NOTE: CHECK STATIC CONVERGENCE AND CONVERGE IF NECESSARY.

1. Reduce blue and green G2 controls to minimum.

2. Loosen front and rear yoke mounting clamp.

3. Slide yoke forward or backward to obtain a red ball on CRT.

4. Spread ears of purity ring only as much as necessary, and rotate complete assembly to center red ball on CRT screen.

5. Position yoke for most uniform red field. Adjust purity ring as necessary to obtain red field.

6. Turn down red G2 control and alternately turn up green G2 and blue G2 controls. Observe individual fields, adjust purity ring as necessary to obtain pure individual fields. In some cases it may be necessary to re-adjust yoke to optimize

DOUBLE SIDED TAPE

FRONT YOKE MTG CLAMP
REAR YOKE MTG CLAMP
GREEN STATIC MAGNET
PURITY RING
BLUE LATERAL MAGNET
UNLOCK BEFORE ADJUSTING BLUE LATERAL

RED STATIC MAGNET

MTG EARS
BLUE STATIC MAGNET

CENTER CORES OVER POLE PIECES
POSITION PURITY RING BETWEEN G1 & G2

CONVERGENCE POLE PIECE ASSEM
APPROX 1/8"
REAR OF CONVERGENCE ASSEMBLY

G2   G1

COLOR TUBE ASSEMBLY AND ADJUSTMENTS

purity of individual red, blue and green fields.

RED BALL

CHECK STATIC CONVERGENCE: reset if necessary. Recheck individual fields. Secure deflection yoke clamps. Proceed with background adjustment.

ASSEMBLING CRT

DEFLECTION YOKE MOUNTING

The following procedure is essential to assure accurate lateral yoke movement necessary to obtain optimum purity.

1. Position yoke firmly against CRT bell and secure rear yoke clamp.

2. Position front yoke mounting clamp onto front of yoke. Snug tighten clamp onto yoke. Slide clamp onto CRT bell and form clamp ears to bell of CRT.

3. Loosen rear yoke clamp to allow yoke movement for purity adjustment.

4. Tighten front and rear yoke clamps after tilt and purity adjustment.

CONVERGENCE YOKE AND BLUE LATERAL ASSEMBLY MOUNTING

1. Install yoke with cores centered over convergence pole pieces as shown in detail, and secure.

2. Install blue lateral assembly with respect to G1 & G2 as shown in detail and secure.

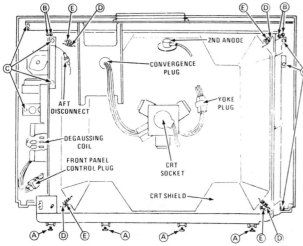

CHASSIS REMOVAL

1. Remove tuner and volume knobs.

2. Unplug yoke, convergence panel, degaussing coil leads, CRT socket, second anode, front panel control plug, AFT disconnect, and ground clips.

3. Place receiver face down on a soft cloth and loosen 4 screws A at bottom of chassis. Perform following steps with receiver face-down.

4. Remove 4 screws B at ends of handle support bracket. Bracket is secured to cabinet by handle studs and is not removed with chassis.

5. Remove 7 screws C at sides of chassis and lift chassis off cabinet front.

CRT REMOVAL

1. Unlock 4 shield springs D and remove shield and degaussing coil as an assembly.

2. Remove 4 bolts E and remove CRT.

Fig. 2-45. CTV6 purity, convergence and CRT replacement instructions.

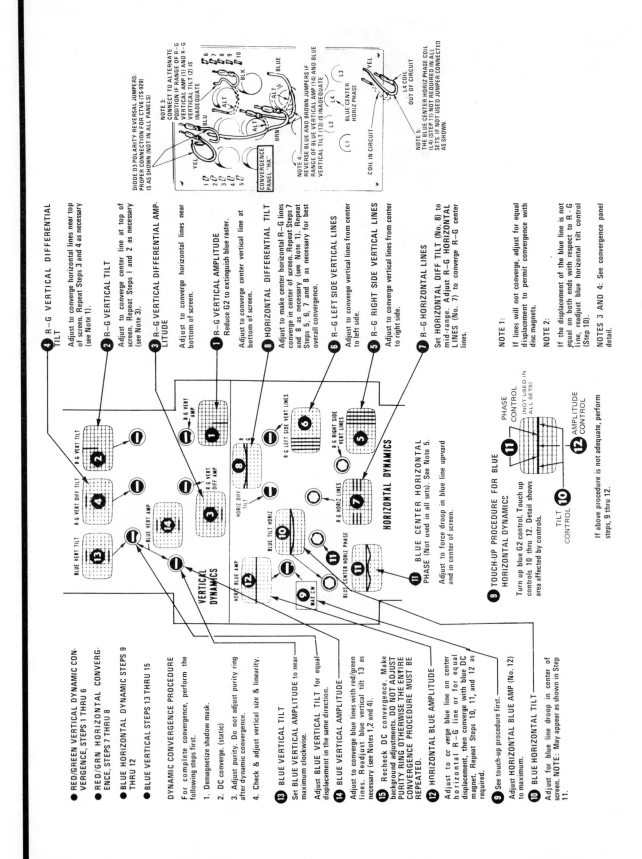

Fig. 2-46. CTV6 dynamic convergence procedure.

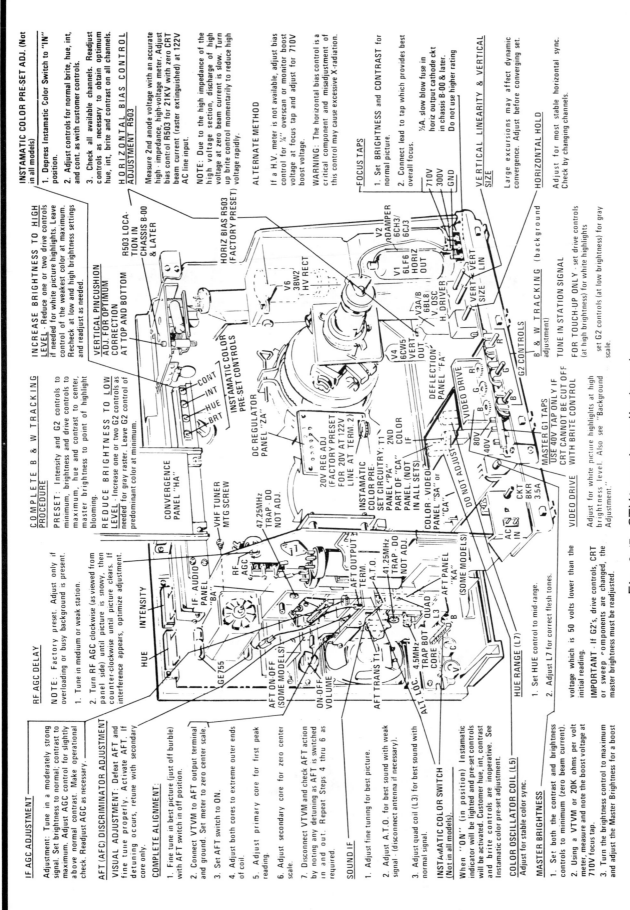

IF AGC ADJUSTMENT

Adjustment: Tune in a moderately strong signal. Set brightness to normal; contrast to maximum. Adjust AGC control for slightly above normal contrast. Make operational check. Readjust AGC as necessary.

AFT (AFC) DISCRIMINATOR ADJUSTMENT

VISUAL ADJUSTMENT: Defeat AFT and fine tune properly. Activate AFT. If detuning occurs, retune with secondary core only.

COMPLETE ALIGNMENT:
1. Fine tune in best picture (just off burble) with AFT switch in off position.
2. Connect VTVM to AFT output terminal and ground. Set meter to zero center scale.
3. Set AFT switch to ON.
4. Adjust both cores to extreme outer ends of coil.
5. Adjust primary core for first peak reading
6. Adjust secondary core for zero center scale.
7. Disconnect VTVM and check AFT action by noting any detuning as AFT is switched in and out. Repeat Steps 4 thru 6 as required.

SOUND IF
1. Adjust fine tuning for best picture.
2. Adjust A.T.O. for best sound with weak signal - (disconnect antenna if necessary).
3. Adjust quad coil (L3) for best sound with normal signal.

INSTAMATIC COLOR SWITCH (Not in all models)
When "ON" (in position) Instamatic indicator will be lighted and pre-set controls will be activated. Customer hue, int, contrast and brite controls are inoperative. See Instamatic color pre-set adjustment.

COLOR OSCILLATOR COIL (L5)
Adjust for stable color sync.

MASTER BRIGHTNESS
1. Set both the contrast and brightness controls to minimum (zero beam current).
2. Using a VTVM or 20K ohms per volt meter, measure and note the boost voltage at 710V focus tap.
3. Turn the brightness control to maximum and adjust the Master Brightness for a boost

RF AGC DELAY
NOTE: Factory preset. Adjust only if overloading or busy background is present.
1. Tune in medium or weak station.
2. Turn RF AGC clockwise (as viewed from panel side) until picture is snowy, then counter-clockwise until picture clears. If interference appears, optimize adjustment.

COMPLETE B & W TRACKING PROCEDURE
PRESET: Intensity and G2 controls to minimum, brightness and drive controls to maximum, hue and contrast to center, master brightness to point of highlight blooming.

REDUCE BRIGHTNESS TO LOW LEVEL - Increase one or two G2 controls as needed for gray raster. Leave G2 control of predominant color at minimum.

INCREASE BRIGHTNESS TO HIGH LEVEL - Reduce one or two drive controls if needed for white picture highlights. Leave control of the weakest color at maximum. Recheck at low and high brightness settings and readjust as needed.

VERTICAL PINCUSHION ADJ. FOR OPTIMUM CORRECTION AT TOP AND BOTTOM

INSTAMATIC COLOR PRE-SET ADJ. (Not in all models)
1. Depress Instamatic Color Switch to "IN" position.
2. Adjust controls for normal brite, hue, int, and cont. as with customer controls.
3. Check all available channels. Readjust controls as necessary to obtain optimum hue, int, brite and contrast on all channels.

HORIZONTAL BIAS CONTROL ADJUSTMENT R503
Measure 2nd anode voltage with an accurate high - impedance, high-voltage meter. Adjust bias control R503 for 21KV with zero CRT beam current (raster extinguished) at 122V AC line input.
NOTE: Due to the high impedance of the high voltage section, discharge of high voltage at zero beam current is slow. Turn up brite control momentarily to reduce high voltage rapidly.

ALTERNATE METHOD
If a H.V. meter is not available, adjust bias control for ¼" overscan or monitor boost voltage at focus tap and adjust for 710V boost voltage.
WARNING: The horizontal bias control is a critical component and misadjustment of this control may cause excessive X-radiation.

FOCUS TAPS
1. Set BRIGHTNESS and CONTRAST for normal picture.
2. Connect lead to tap which provides best overall focus.
710V
300V
GND

VERTICAL LINEARITY & VERTICAL SIZE
Large excursions may affect dynamic convergence. Adjust before converging set.

HORIZONTAL HOLD
Adjust for most stable horizontal sync. Check by changing channels.

HUE RANGE (L7)
1. Set HUE control to mid-range.
2. Adjust L7 for correct flesh tones.

voltage which is 50 volts lower than the initial reading.
IMPORTANT - If G2's, drive controls or sweep components are changed, the master brightness must be readjusted.

VIDEO DRIVE: Adjust for white picture highlights at high brightness level. Also see "Background Adjustment."

Diagram labels: R503 LOCATION IN CHASSIS B-00 & LATER; HORIZ BIAS R503 (FACTORY PRESET); V6 3BW2 HV RECT; V2 6CH3/6CJ3 DAMPER; V1 6LF6 HORIZ OUT; V3A/B 6BL8 V. OSC H. DRIVER; V4 6CW5 VERT OUT; VERT VERT SIZE LIN; G2 CONTROLS; B' & W TRACKING (background adjustment); TUNE IN STATION SIGNAL; FOR TOUCH-UP ONLY - set drive controls (at high brightness) for white highlights set G2 controls (at low brightness) for gray scale.; VIDEO DRIVE R G B; MASTER G1 TAPS USE 40V TAP ONLY IF CRT CANNOT BE CUT OFF WITH BRITE CONTROL; 80V 40V; DO NOT ADJUST; COLOR - VIDEO PANEL "SA" or "CA"; INSTAMATIC COLOR PRE-SET CIRCUITRY: T1 PANEL "PA" PART OF "CA" PANEL (NOT IN ALL SETS); 20V REG ADJ (FACTORY PRESET FOR 20V AT 122V LINE AT TERM. 2); 2ND COLOR IF; DC REGULATOR PANEL "ZA"; CONVERGENCE PANEL "HA"; VHF TUNER MTG SCREW; 41.25MHz TRAP - DO NOT ADJ.; CONT INT HUE BRT; INSTAMATIC COLOR PRE-SET CONTROLS; AC IN; CKT BKR 3.5A; AFT PANEL "KA" (SOME MODELS); AFT OUTPUT TERM.; ALT. LOC. 4.5MHz TRAP BOT QUAD CORE L3; AFT TRANS T1; AFT ON-OFF (SOME MODELS); ON-OFF VOLUME; HUE; INTENSITY; RF AGC; IF- AUDIO PANEL "BA"; GE755.

¼A. slow blow fuse in horiz output cathode ckt in chassis B-00 & later. Do not use higher rating

Fig. 2-47. CTV6 service adjustments.

53

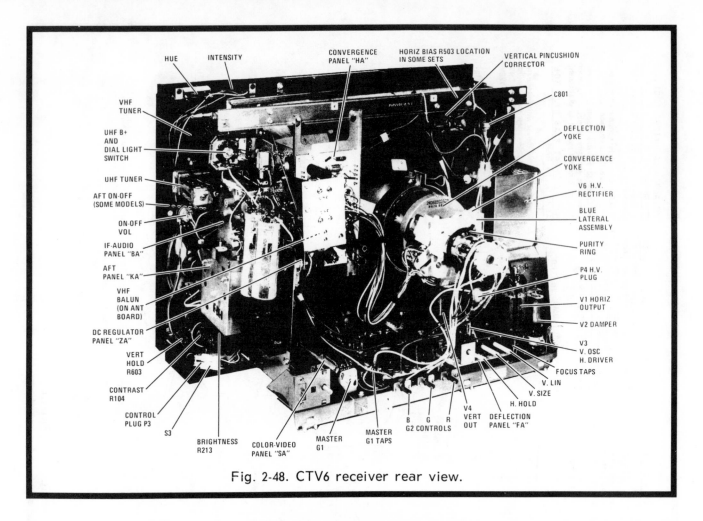

Fig. 2-48. CTV6 receiver rear view.

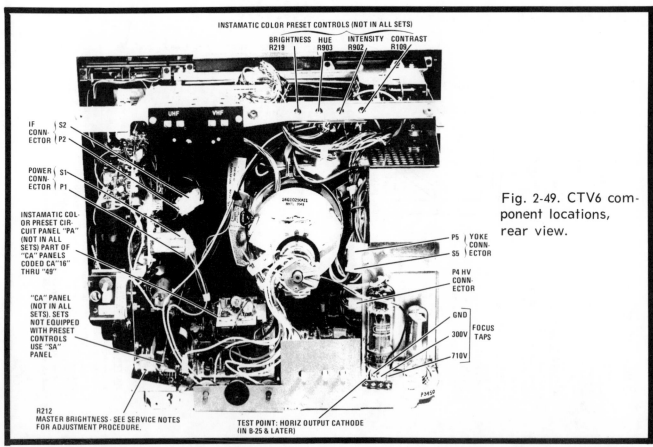

Fig. 2-49. CTV6 component locations, rear view.

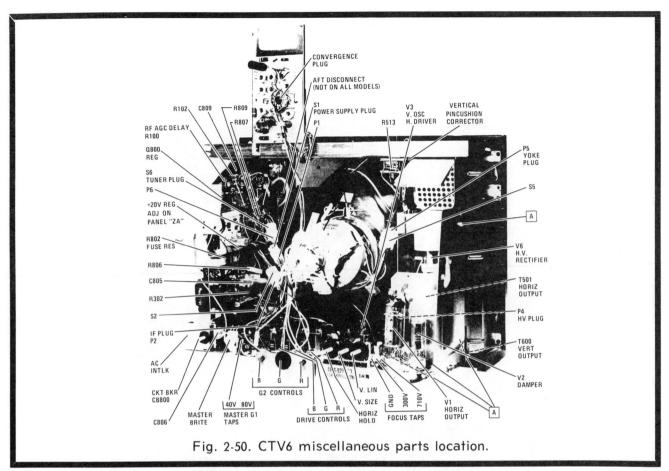

Fig. 2-50. CTV6 miscellaneous parts location.

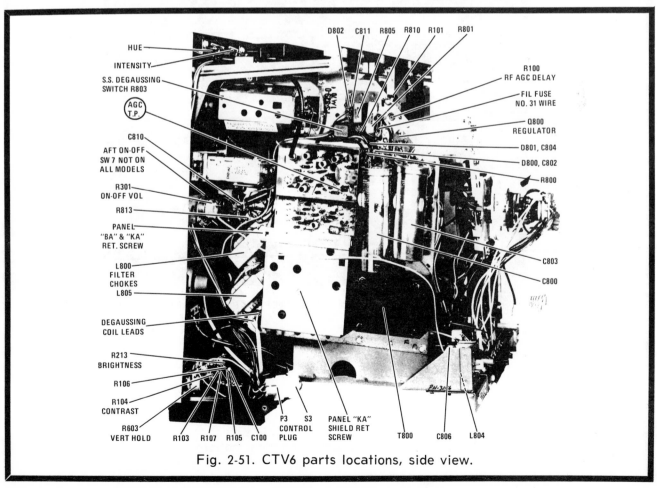

Fig. 2-51. CTV6 parts locations, side view.

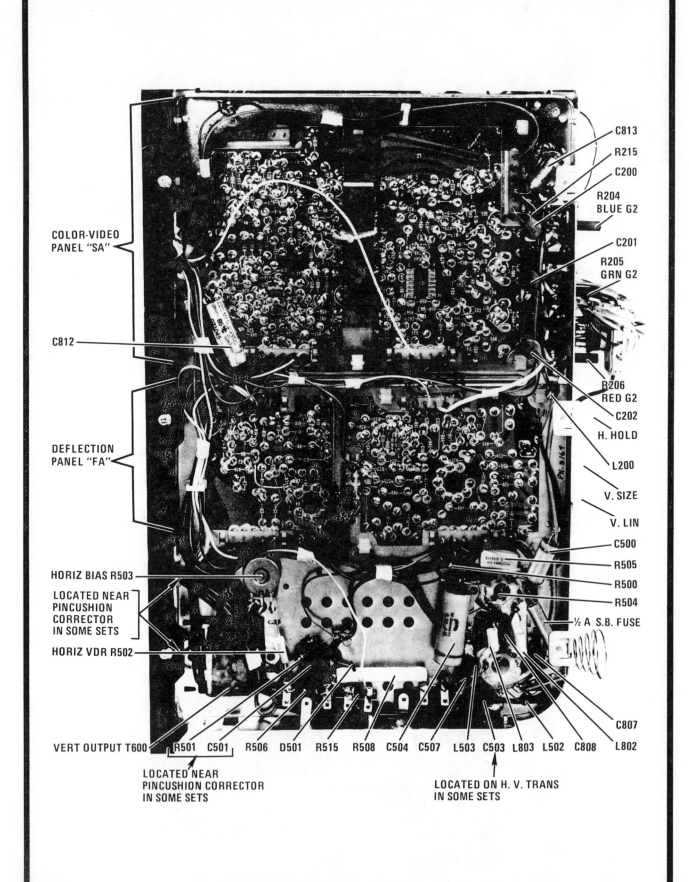

COLOR-VIDEO PANEL "SA"

C812

DEFLECTION PANEL "FA"

HORIZ BIAS R503

LOCATED NEAR PINCUSHION CORRECTOR IN SOME SETS

HORIZ VDR R502

VERT OUTPUT T600

LOCATED NEAR PINCUSHION CORRECTOR IN SOME SETS

R501 C501 R506 D501 R515 R508 C504 C507 L503 C503 L803 L502 C808 L802

LOCATED ON H. V. TRANS IN SOME SETS

C813
R215
C200
R204 BLUE G2
C201
R205 GRN G2
R206 RED G2
C202
H. HOLD
L200
V. SIZE
V. LIN
C500
R505
R500
R504
½ A S.B. FUSE
C807

Fig. 2-52. CTV6 off-panel parts location, bottom view.

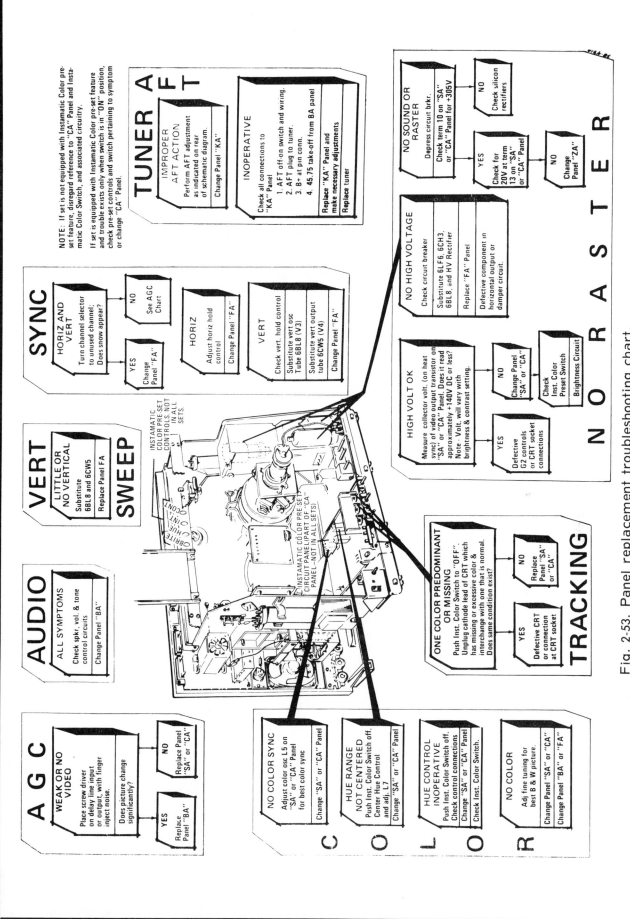

Fig. 2-53. Panel replacement troubleshooting chart.

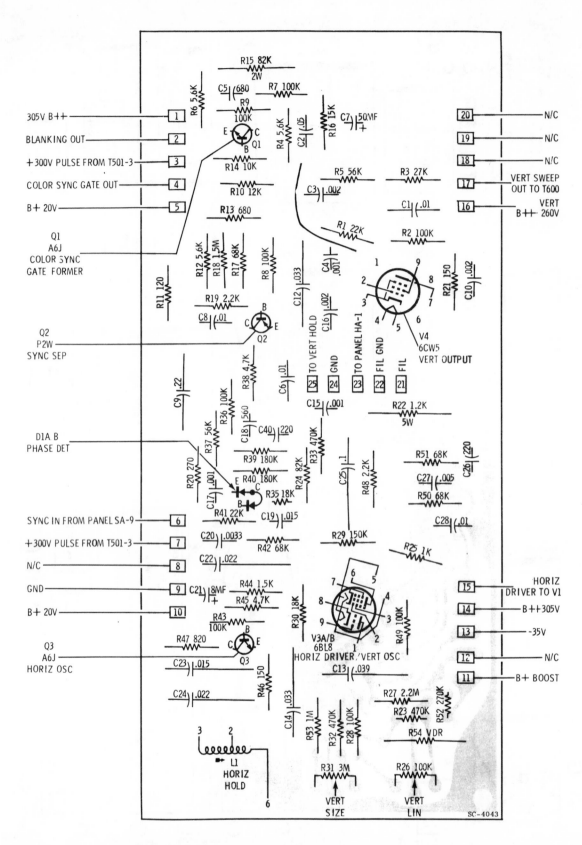

Fig. 2-54. Deflection panel FA, component side.

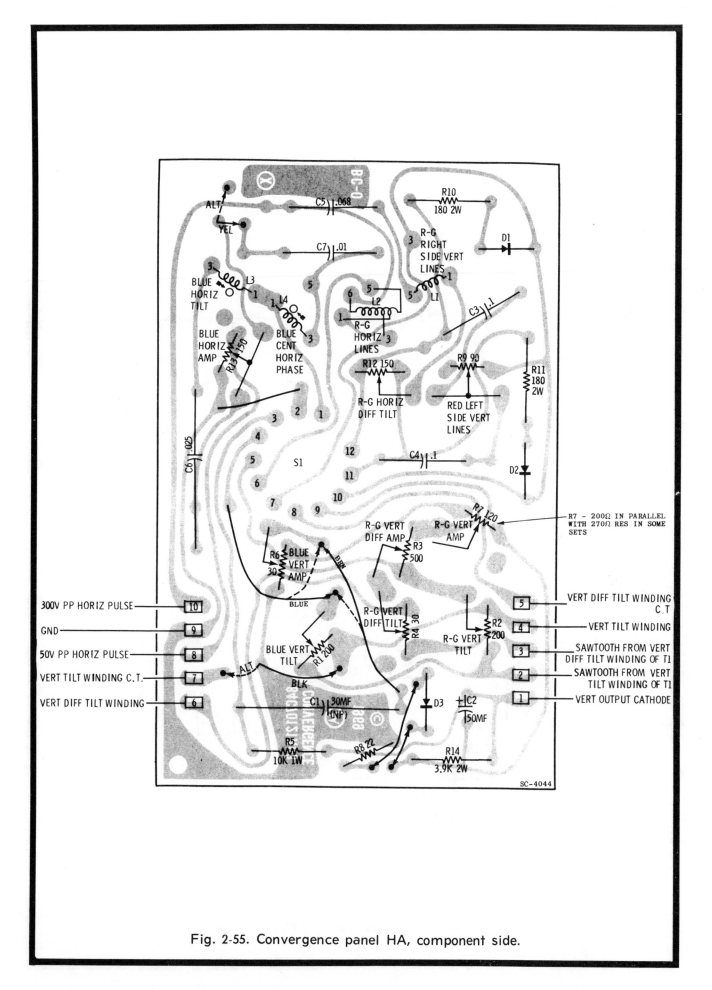

Fig. 2-55. Convergence panel HA, component side.

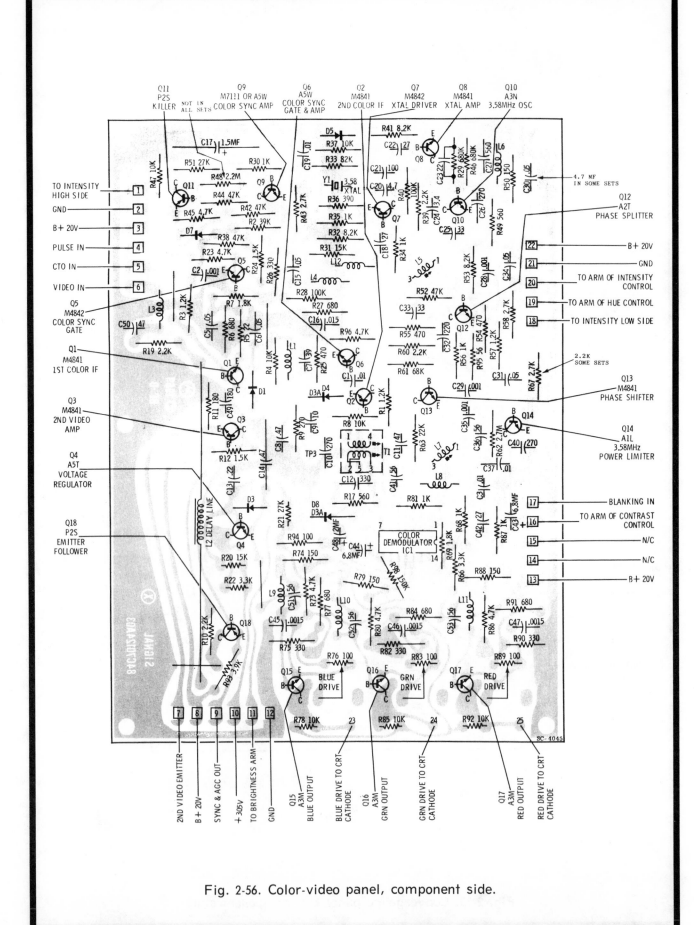

Fig. 2-56. Color-video panel, component side.

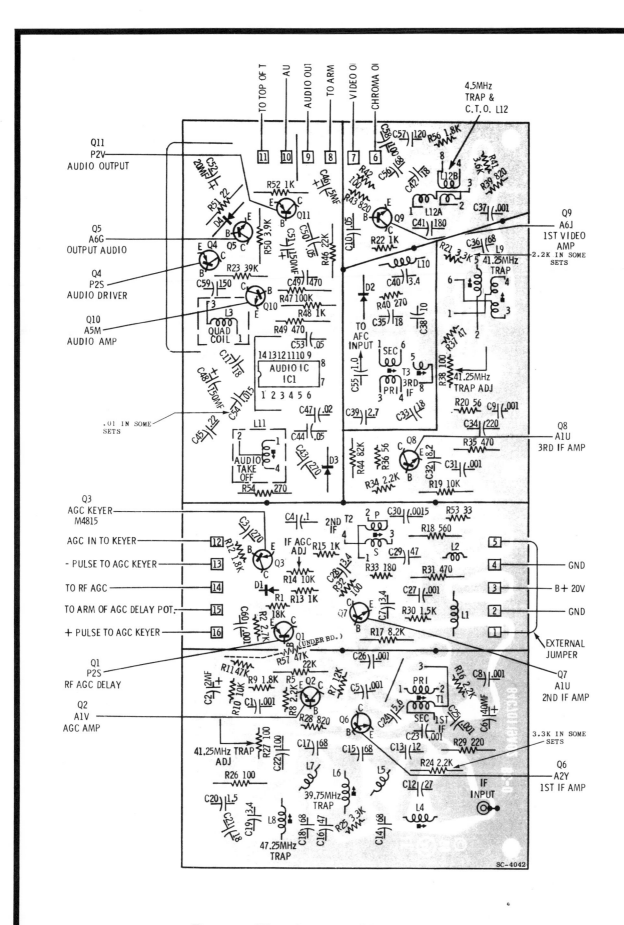

Fig. 2-57. IF audio panel BA, component side.

61

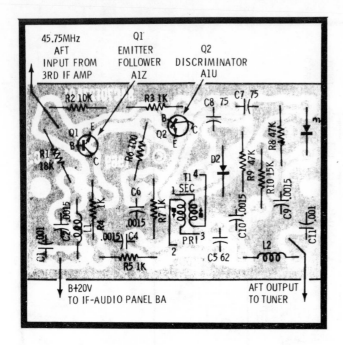

Fig. 2-58. AFT panel KA, component side.

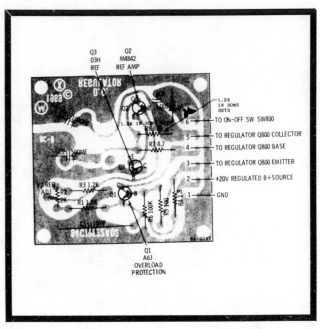

Fig. 2-59. DC regulator panel ZA, component side.

CHAPTER 3

Alignment: New Quasar & Quasar II

With thanks to Herb Bowden and Jim Smith, Sencore, 3200 Sencore Drive, Sioux Falls South Dakota 57107, we present the following simple sweep alignment instructions for Quasar II and new Quasar receivers. This information appears in the **Speed Aligner Workshop Manual** series currently being offered by Sencore representatives throughout the United States to all television service facilities. In these instructions, the SM158 is used exclusively, but the larger and more flexible SM152 (my own sweep-marker generator) can be used exactly as directed also.

All instructions are direct, simple and include an RF-IF overall check, straight IF, trap, AFT, and chroma alignment in much that order. Individual peak-to-peak amplitudes of the various response curves are not given. Therefore, be careful to stay strictly within bias limits and set the scope as generally shown to produce voltages that measure about 2 volts peak-to-peak. Remember that high settings of the crystal markers can produce "noise," so keep marker amplitudes at a reasonable but comfortable level for good viewing and best response-sweep performance.

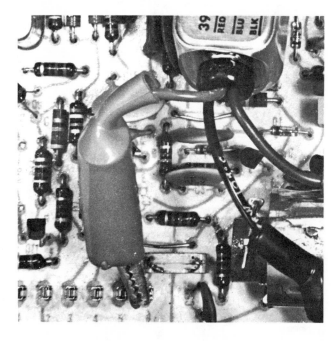

 Connect RF Cable of SM158 to antenna input using special jack provided. (Red clip to jack pin, black to chassis.)

 Connect Detector Probe (Red Lead) to Video Detector test point Terminal 6 on "SA", "CA" OR "TA" panel output 1st video emitter follower.

1. Set SM158 RF output to Channel 10.

2. Set Fine Tuning, Sweep Width, and Sweep Height to midrange.

3. Depress 41.25, 42.17, 45.75, and 47.25MHz markers.

4. Set Market Height control to midrange.

5. Set Chassis tuner to Channel 10.

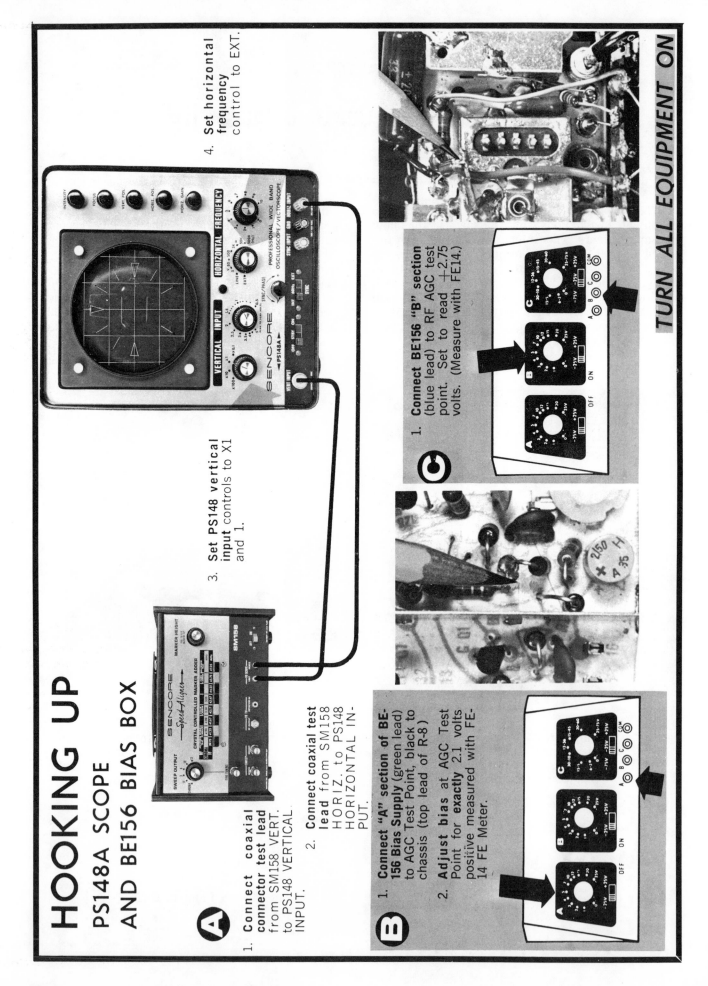

HOOKING UP
PS148A SCOPE AND BE156 BIAS BOX

TURN ALL EQUIPMENT ON

A
1. Connect coaxial connector test lead from SM158 VERT. to PS148 VERTICAL INPUT.
2. Connect coaxial test lead from SM158 HORIZ. to PS148 HORIZONTAL INPUT.

3. Set PS148 vertical input controls to X1 and 1.

4. Set horizontal frequency control to EXT.

B
1. Connect "A" section of BE-156 Bias Supply (green lead) to AGC Test Point, black to chassis (top lead of R-8)
2. Adjust bias at AGC Test Point for exactly 2.1 volts positive measured with FE-14 FE Meter.

C
1. Connect BE156 "B" section (blue lead) to RF AGC test point. Set to read +2.75 volts. (Measure with FE14.)

RF/IF OVERALL CHECK

A

2″ SQ.

Adjust SM158 sweep height so that response curve fills two squares on scope; (2vpp); center response curve with PS148 Horiz. Gain control and SM158 fine tuning and sweep width controls.

B

Color carrier (42.17MHz) and video carrier (45.75MHz) should appear across from each other.

Adjust chassis fine tuning so that 47.25 MHz marker and 41.25 MHz marker are in dips on either side of curve.

C

Switch SM158 and chassis tuner to channels 13, 4 and 3.

If curve is tilted or markers not correctly positioned, slightly readjust L4 and top slug of T3 to shape curve and position markers.

I. F. ALIGNMENT
ALIGNING 3RD I. F. STAGE
HOOK-UP

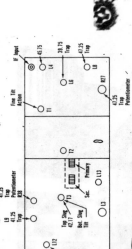
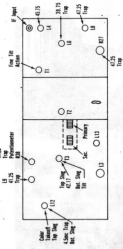

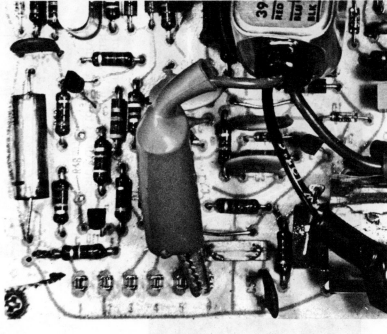

R32 Exposed Lead
"BA" Panel IF

A **Connect RF cable and matching pad to test point 1** (collector of 2nd IF). Red lead to test point (exposed lead of R32) and black to chassis.

All other connections remain the same.

B **Set Sweep Output control to IF.** Adjust Sweep Height control for 2 inch response (2vpp).

C Release all markers except the 41.25 MHz trap marker.

I. F. ALIGNMENT
ALIGNING 3RD I. F. STAGE

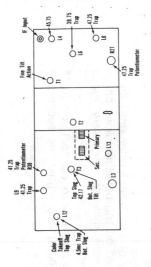

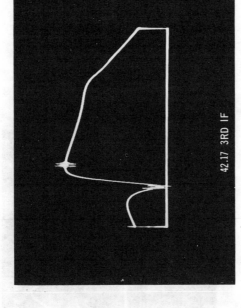

42.17 3RD IF

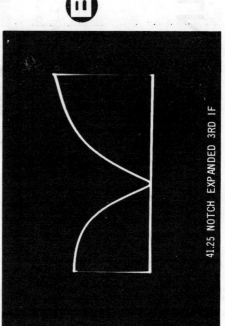

41.25 EXPANDED 3RD IF

A Adjust **L9, 41.25MHz trap** to place trap notch on the 41.25MHz marker. (Action of trap more easily seen if response curve is expanded by reducing sweep width and increasing sweep height on the SM158.)

B Adjust **R38, 41.25MHz trap** pot to place trap notch as close to baseline as possible. (Reducing marker amplitude will make trap notch more visible. In some cases, trap notch will extend below baseline. In this event, adjust trap notch as far below baseline as possible.) After adjusting R38, increase marker height and readjust L9 to place trap notch on marker.

41.25 NOTCH EXPANDED 3RD IF

C Depress 42.17MHz marker.

Adjust **top** slug of 3rd IF (T3) transformer to place 42.17MHz marker as far from baseline as possible.

Adjust **bottom** slug of T3 to place 42.17MHz marker as far from baseline as possible **without shifting 41.25MHz marker.**

Recheck setting of 41.25MHz trap.

I.F. ALIGNMENT
AFT ADJUSTMENT

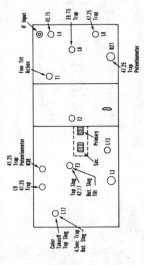

A

1. Connect detector probe to AFT test point. (Red lead to AFT output, black to chassis.)

2. Disconnect AFT output from tuner.

3. All other connections remain the same.

4. Adjust PS148 Oscilloscope vertical input to .2vpp/in. X 0.1 and 2)
   Adjust SM158 RF output for 2 inch response.

5. Depress 45.75 marker.

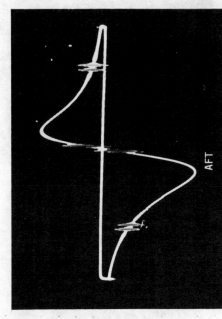

B

1. Adjust primary of AFT transformer (first core to edge of can) for a symmetrical curve with maximum amplitude. Curve will be about three units above the baseline and four units below at best setting.

3 units		.15v
4 units		.2v

2. Adjust secondary of AFT transformer until 45.75 marker is centered on the baseline. A touch-up of the primary core may be necessary for a symmetrical curve.

   An alternate procedure is to connect the FE14 to the AFT test point and adjust the secondary for zero volts.

I.F. ALIGNMENT

ADJACENT SOUND TRAP ADJUSTMENT

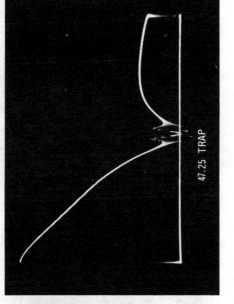

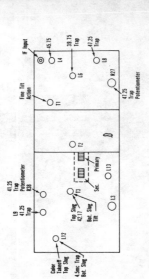

TEST PT. M ON TUNER

 A

1. Connect SM158 RF cable and matching pad to Mixer test point M. (red lead to test point, black to chassis. Shown for TS-934).

2. Re-connect Detector Probe (Red Lead) to Video Detector Test Point 6 "SA".

3. Reset PS148 Oscilloscope vertical input controls to 1vpp/in. (X1 & 1)

B

1. Depress 47.25 MHz marker.

2. Adjust L8, 47.25MHz trap to place trap notch on marker. (Action of trap is more easily seen if trace is expanded by reducing sweep width and increasing sweep height.)

C

1. Adjust R27, 47.25MHz trap pot to place trap notch as close to baseline as possible. (Reducing marker amplitude will make the trap notch more visible. In some cases, trap notch will extend below baseline. In this event, adjust trap notch as far below baseline as possible.)

2. After adjusting R27, increase marker height and readjust L8 to place trap notch on marker.

I.F. ALIGNMENT

39.75 ADJACENT VIDEO TRAP & OVERALL
PRELIMINARY ADJUSTMENT

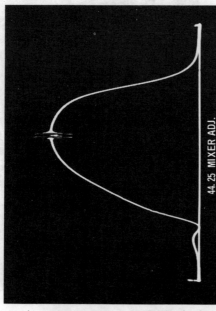

44.25 MIXER ADJ.

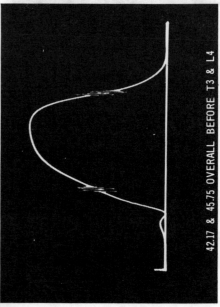

42.17 & 45.75 OVERALL BEFORE T3 & L4

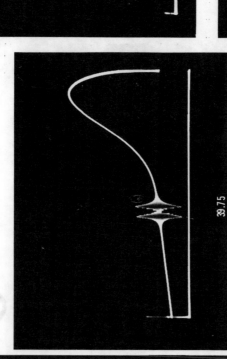

39.75

A

1. Depress 39.75MHz marker and release all others.

2. Adjust L6, 39.75MHz trap to place
   39.75MHz marker as close to
   baseline as possible.

   (Trap action is more easily seen
   if the response curve is expanded by reducing sweep width and
   increasing sweep height on the
   SM158.)

B

1. Readjust SM158 controls for 2"
   response curve.

2. Depress 44.25MHz marker.

3. Adjust L23 mixer coil in tuner to
   place 44.25MHz marker as far
   from baseline as possible.

C

1. Depress 45.75MHz marker.

2. Adjust L4, 1st IF input coil, to
   place 45.75MHz marker at 50%
   response.

3. Depress 42.17MHz marker.

4. Adjust top slug T3, 3rd IF transformer, to place 42.17MHz marker
   at 50% response.

I.F. ALIGNMENT
OVERALL IF ALIGNMENT

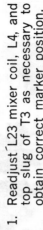

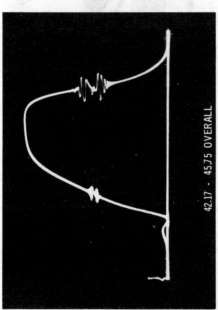

42.17 - 45.75 OVERALL

B

1. Readjust L23 mixer coil, L4, and top slug of T3 as necessary to obtain correct marker position.

2. For ease in positioning markers, pull out on Marker Height switch to obtain horizontal markers as shown.

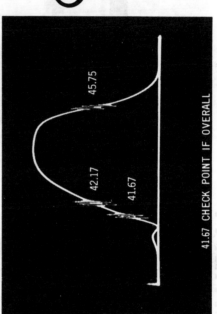

41.67 CHECK POINT IF OVERALL

C

1. Depress 41.67MHz marker.

2. It must be no less than 15% amplitude on response curve.

3. If marker is low, slightly readjust top slug of T3 to raise marker to 15%.

4. If 15% cannot be obtained, recheck the setting of the 41.25 MHz trap.

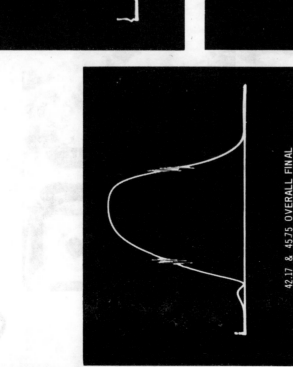

42.17 & 45.75 OVERALL FINAL

A

1. Adjust T1, 1st IF transformer, and T2, 2nd IF transformer, to remove tilt and obtain maximum amplitude.

(These adjustments may not cause as great a response change as other adjustments. This is a normal action.)

CHROMA ALIGNMENT
HOOKING UP

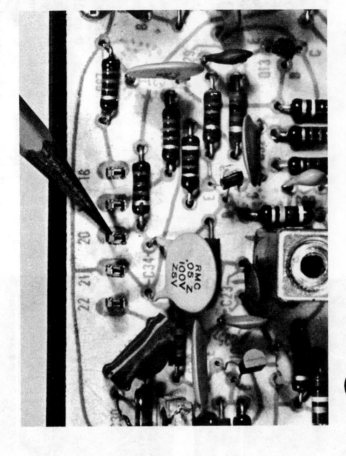

B

1. Connect "C" section of BE156 (red lead) to terminal 20 "SA". Adjust "C" section of BE156 for 2 volts positive measured at 20 "SA".

C

1. Set PS148 vertical input controls to .25 vpp/in. (X.1 and 2.5)

2. Set SM158 sweep output to Chroma sweep. Depress 4.5, 4.08, and 3.08MHz markers. Adjust sweep height for a response of 2". (.5vpp)

A

1. Connect SM158 detector cable blue demodulator lead to pin 6 of IC demodulator. (Use lead of R17 as shown.) Connect black lead to chassis ground.

2. All other SM158 and PS148 connections remain the same as in the previous step.

3. Adjust "A" section of BE156 to obtain +2.5 volts measured at IF AGC test point.

CHROMA ALIGNMENT

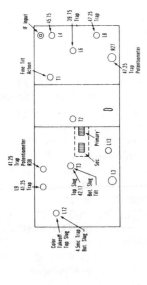

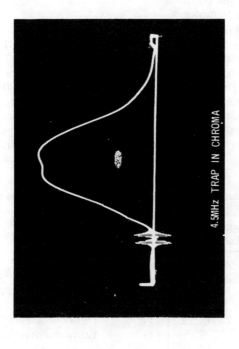

4.5MHz TRAP IN CHROMA

A 1. Adjust L12, bottom slug, 4.5MHz trap for minimum response at 4.5MHz.

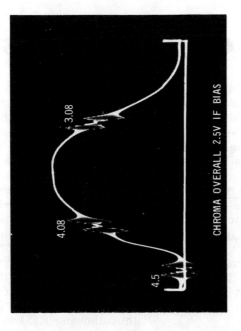

CHROMA OVERALL 2.5V IF BIAS

B
1. Adjust L12, color take-off coil, top slug, for maximum amplitude.

2. Adjust L12, bottom slug, 4.5MHz trap to position the 4.5MHz marker as close to baseline as possible.

C
1. Adjust L12, top slug, color take-off coil, to place 4.08MHz marker as far from baseline as possible.

2. Adjust T1, bandpass transformer, (on "SA" Color/Video Panel) top and bottom slugs for correct marker position and curve symmetry.

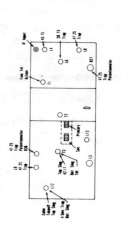

CHAPTER 4

Troubleshooting the Original Quasar

Since the original Quasar is covered in the **Motorola Color TV Service Manual, Vol. 1** (TAB 509) and in **Modern TV Circuit & Waveform Analysis** (TAB 476), here we concentrate on trouble case histories and modifications. Listed is all available information on each plug-board, along with the various modifications so you may determine immediately if the panel or chassis you're servicing has either the latest panel coding or most recent panel coding changes. Again, we suggest you turn in all your panels once a year, if they aren't the most recent, and use new units for best servicing and customer satisfaction. After all, even the retail prices aren't exhorbitant, as exhibited in Table 4-1, and with dealer terms they should continue quite attractive.

AC LINE REGULATOR

The most important circuit revision in the newer Quasar solid-state TS-915 series was made in the AC-regulated power supply (Fig. 4-1). Fig. 4-2 is a schematic of the regulating circuits. There is only one adjustment, and this is normally **not** required in the usual home setup. If, however, you have to make a regulator adjustment, you must **readjust** the **high voltage** and **width** also.

To set the regulator, turn the brightness and contrast controls to minimum for zero beam current, connect an accurate voltmeter to Terminal 6Z on the panel, and, with an insulated tool, adjust regulator control R5Z for 105 volts DC. (This should insure AC regulation from some 105 to 130 volts.) To set the high voltage, adjust the HV control on pincushion panel G for a slight overscan of the horizontal raster width.

WARNING: High current AC and DC voltages are both present on the regulator panel. For your own safety and that of the receiver, be especially careful when using either a voltmeter or an oscilloscope in this section of the receiver; the probes might cause an unpleasant short.

With a high AC line condition, the B+ at Point 6Z (Fig. 4-2) rises and Q1Z conducts less current. C3Z takes longer to charge (as compared to a low B+ condition) through Q1Z. When the charge on C3Z reaches 28v, T3Z is fired. The C3Z discharge pulse triggers on E1Z, which shorts R15J and R16J. EZ1 remains on until the AC across A1 and A2 goes to zero; when this occurs, the cycle repeats itself. Because E1Z is fired late, its duty cycle is short, thus allowing R15J and R16J to be in the circuit longer during each AC alternation. This counteracts the original rise in the AC line. Q2Z is pulsed on at the beginning of each AC alternation to establish a zero reference. With a low AC line condition, the exact reverse of the above conditions occurs.

Servicing the Regulator

All components and circuitry, except for the two series-connected 50-watt resistors (R15J and R16J) are located on panel Z. Thus, regulator troubles can be isolated to either panel Z or the resistors.

The circuits within the heavy line in Fig. 4-2 are located on the panel. When the panel is removed, AC is still applied to the power and filament transformers through the two 50-watt resistors. If the receiver is operated without the panel, a reduction in raster size is likely because the B+ voltage is lower.

Quasar COLOR TV			Quasar II COLOR TV		
Panel	Description	Suggested List Exchange*	Panel	Description	Suggested List Exchange*
B	IF Panel	$12.50	SA	Color & Video Panel	$18.00
P	FT1 Panel	6.25	BA	IF & Audio Panel	14.00
D	IC Audio Panel	7.50	ZA	DC Regulator Panel	6.50
E	Video Pre-Amp Panel	11.25	KA	AFC Panel	7.50
S	Color Panel	12.50	FA	Deflection Panel	11.00
L	Video Drive Panel	8.75	HA	Convergence Panel	10.00
M	Video Output Panel	10.00			
F	Horizontal Panel	14.00			
G	Pincushion Panel	7.50			
H	Convergence Panel	13.75			
T	FTL Panel	7.50	*Manufacturer's Suggested Prices. Optional with dealer.		

Table 4-1. "Works in a drawer" field replacement panels.

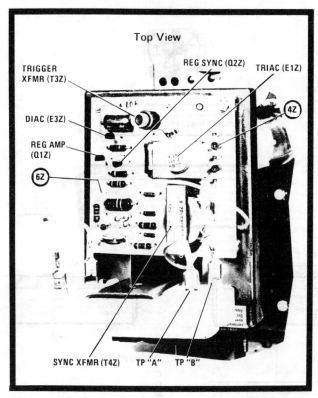

Top View

TRIGGER XFMR (T3Z)

REG SYNC (Q2Z)

TRIAC (E1Z)

DIAC (E3Z)

REG AMP (Q1Z)

6Z

4Z

SYNC XFMR (T4Z)   TP "A"   TP "B"

Fig. 4-1. AC line regulator and power supply.

Before the panel is condemned, be certain that the B+ source voltages are present. At 6Z you should read +105v and at 4Z +100v. Be extremely careful with test prods if you measure the voltage at 4Z.

In the step-by-step charts, refer to the chart corresponding to the symptom and follow the suggested action in each block.

UHF AUTOMATIC FINE TUNING, DD25TS-915 & C25TS-915

A temperature-compensated zener diode, D6A in Fig. 4-3, produces a 33-volt regulated output from a 95-volt source. Thirteen variable resistors are placed across the 33-volt supply. To maintain constant loading and minimize drift, the resistors are always in the circuit. A wiper contact removes the voltage from the arm of the variable resistor by means of a contact (one of 13) switched in the circuit for a specific channel. The 33 volts is applied to four varicap diodes, reverse biasing them. The resultant capacity tunes the circuit to a specific channel.

Automatic fine tuning, applied to the oscillator in the VHF tuner, is also applied to the UHF tuner. The AFT defeat switch, when activated, grounds the AFT voltage, allowing the UHF tuner to operate without external control.

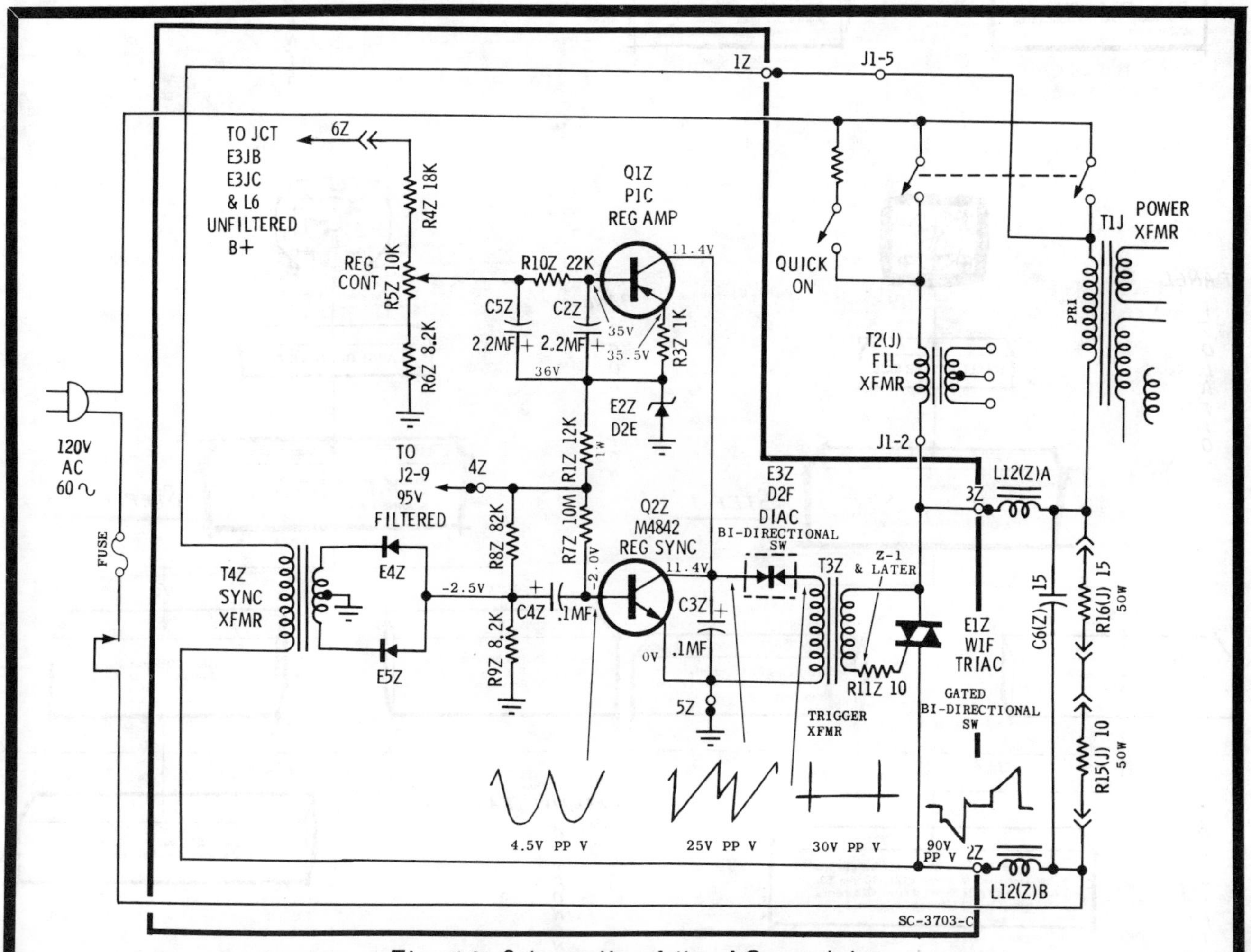

Fig. 4-2. Schematic of the AC regulator.

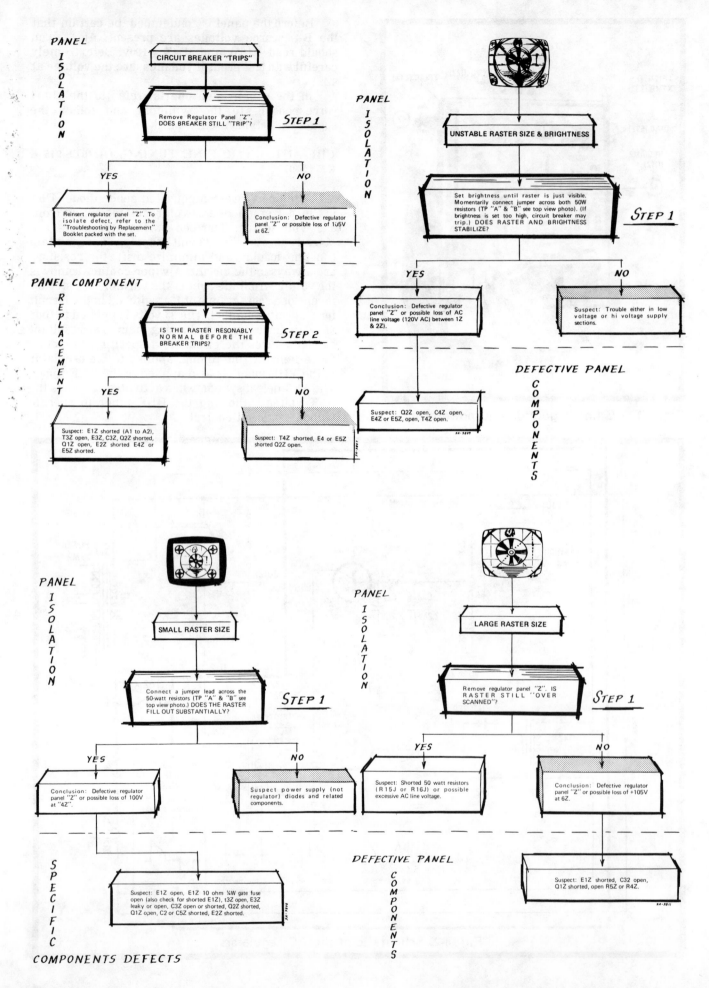

CIRCUIT BREAKER "TRIPS"

Remove Regulator Panel "Z". DOES BREAKER STILL "TRIP"? *STEP 1*

YES

Reinsert regulator panel "Z". To isolate defect, refer to the "Troubleshooting by Replacement" Booklet packed with the set.

NO

Conclusion: Defective regulator panel "Z" or possible loss of 105V at 6Z.

IS THE RASTER RESONABLY NORMAL BEFORE THE BREAKER TRIPS? *STEP 2*

YES

Suspect: E1Z shorted (A1 to A2), T3Z open, E3Z, C3Z, Q2Z shorted, Q1Z open, E2Z shorted E4Z or E5Z shorted.

NO

Suspect: T4Z shorted, E4 or E5Z shorted Q2Z open.

UNSTABLE RASTER SIZE & BRIGHTNESS

Set brightness until raster is just visible. Momentarily connect jumper across both 50W resistors (TP "A" & "B" see top view photo). (If brightness is set too high, circuit breaker may trip.) DOES RASTER AND BRIGHTNESS STABILIZE? *STEP 1*

YES

Conclusion: Defective regulator panel "Z" or possible loss of AC line voltage (120V AC) between 1Z & 2Z).

NO

Suspect: Trouble either in low voltage or hi voltage supply sections.

Suspect: Q2Z open, C4Z open, E4Z or E5Z, open, T4Z open.

SMALL RASTER SIZE

Connect a jumper lead across the 50-watt resistors (TP "A" & "B" see top view photo.) DOES THE RASTER FILL OUT SUBSTANTIALLY? *STEP 1*

YES

Conclusion: Defective regulator panel "Z" or possible loss of 100V at "4Z".

NO

Suspect power supply (not regulator) diodes and related components.

Suspect: E1Z open, E1Z 10 ohm ¼W gate fuse open (also check for shorted E1Z), t3Z open, E3Z leaky or open, C3Z open or shorted, Q2Z shorted, Q1Z open, C2 or C5Z shorted, E2Z shorted.

LARGE RASTER SIZE

Remove regulator panel "Z". IS RASTER STILL "OVER SCANNED"? *STEP 1*

YES

Suspect: Shorted 50 watt resistors (R15J or R16J) or possible excessive AC line voltage.

NO

Conclusion: Defective regulator panel "Z" or possible loss of +105V at 6Z.

Suspect: E1Z shorted, C32 open, Q1Z shorted, open R5Z or R4Z.

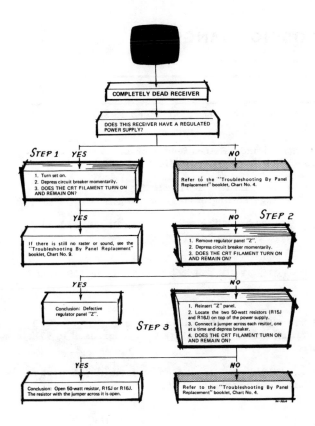

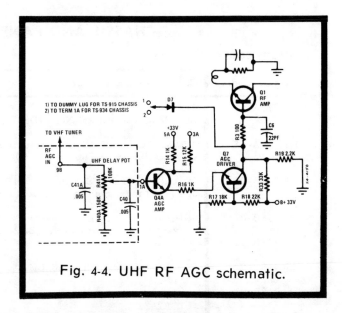

Fig. 4-4. UHF RF AGC schematic.

The AFT voltage (approximately 2 volts) is divided by R37 and R38, and about 10 percent is applied to the gate of an FET. The output voltage, taken from the drain, is DC-coupled by R21 to the varicap control voltage. If a change in the AFC correction occurs, a corresponding change in DC voltage to the varicap diodes tunes the circuits and maintains the proper frequency. The DC tuning control voltage will vary from 1.2 to 34.5 volts at the varicap diodes.

UHF RF AGC

When a signal of sufficient strength is received, the voltage at Terminal 9B in Fig. 4-4 rises and the

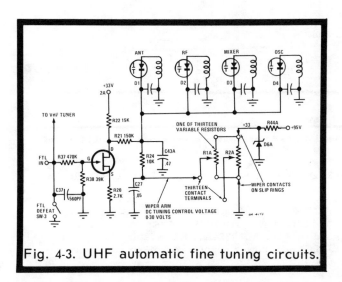

Fig. 4-3. UHF automatic fine tuning circuits.

setting of the RF AGC control for UHF determines the voltage applied to Q4A, an emitter-follower stage. The AGC amplifier collector circuit is in a feedback loop with the video IF AGC. This loop determines when the RF amplifier will begin reducing gain.

The output of Q4A is applied to Q7A (AGC driver). This grounded-base amplifier controls the amount of forward bias applied to the base of the RF amplifier, Q1A. Positive voltage (about 2.5 volts) is applied to the base of the RF amplifier for maximum gain. With stronger signals, Q7A increases in conduction, resulting in an increase in forward bias to the RF amplifier and a reduction in gain.

TS915-919 Case Histories

Video IF **Panel B** has different AGC connections, depending on the series, and the wrong connections can cut out the high channels, make other intermittent, plus give every indication of a faulty tuner. Beware!

Remote problems can result from arcovers across excess solder on remote panels to other points on the chassis, usually around panel switches, and will cause pops in audio, even color, hue, and volume changes. A new IC panel (D-12) can also remove audio problems.

Air bubble in the silicon high-voltage "tire" cover causes a corona arc that can damage both transformer and surrounding parts.

No sound, no video sometimes is caused when the FTI blue wire is pinched.

E4J half-wave rectifier in the 240-volt power supply can present several problems. If it opens, the symptoms look like a gassy picture tube. If it shorts, there are heavy retrace lines, and the circuitbreaker pops repeatedly.

CHASSIS AND PANEL CODING CHANGES

CHASSIS CODING CHANGES

Chassis Coding	Chassis Coding Changes
TS-919A-02	TO REDUCE UHF REGENERATION: A 7.5uh choke (L-12A) added in series with RF AGC lead, choke Part No. 24D66772A12.
TS-919A-08	TO IMPROVE RF AGC: Resistor R-50A (10K) and RF AGC feed back line added between IF AGC line and RF amplifier (Q-1) collector load (see schematic diagram).
TS-915B-00	INCORPORATES TS-919A-02 THRU A-08 CHANGES IN TS-915 CHASSIS: CRT G-1 voltage (35V) increased to 95V. Requires mid-range setting of G-2 controls.
TS-915B-07 and TS-919B-03	SUPPORT BRACKET ADDED TO HIGH VOLTAGE TRANSFORMER: Bracket added to prevent shipping damage to 3BN2, high voltage rectifier tube.
TS-915B-08 and TS-919B-02	POWER SUPPLY DIODES CHANGED TO IMPROVED TYPE: New Part No. 48S191A07 (E-4J).
C23TS-919B-00	ADDED FINE TUNING LOCK TO TS-919 CHASSIS: Required changing VHF and UHF tuners (see schematic diagram for details). CRT G-1 voltage (35V) increased to 95V. Requires mid-range setting of G-2 controls.
TS-919B-01 and TS-915B-09	TO IMPROVE RELIABILITY OF AUDIO OUTPUT, Q-5D: Added varistor R-31D across primary of audio output transformer. Varistor is Part No. 6C66263A08.
TS-919B-00	TO IMPROVE 35V B+ LINE FILTERING: Add 100mfd, 80V capacitor, Part No. 23C65808A22, from wiring harness terminal 10F to chassis.
TS-919B-09 and TS-915B-00	FILAMENT TAP ADDED TO CRT FILAMENT TRANSFORMER, T-2J: CRT filament transformer changed to tapped type with moveable tap to accommodate different CRT currents. Allows several different CRT's to be used.
TS-919B-10 and TS-915B-12	TO IMPROVE RELIABILITY OF Q-1R, HORIZONTAL REGULATOR TRANSISTOR: Add 1K, 1/2 watt resistor from base to emitter of Q-1R, horizontal regulator transistor.
TS-915B-18 and later, TS-919B-08-6, TS-919B-09-6, TS-919B-10-6 only and TS-919B-16 and later	TO IMPROVE BLANKER AMPLIFIER (Q-6L) AND BLANKER OUTPUT (Q-7L) RELIABILITY: 33K, 10%, 1/2 watt resistor added in base circuit of blanker amplifier, Q-6L. Resistor is mounted on chassis plug between terminals 2F and 5F (terminal 5F was not used previously).
TS-915C-02 and TS-919C-02	TO REDUCE RESIDUAL HUM OR BUZZ WHEN VOLUME IS SET AT LOW LEVEL: Remove ground wire from panel connector to ground. Chassis TS-915 - Terminal 3D and 6D. Chassis TS-919 - Terminal 2D and 6D.

POWER SUPPLY CODING CHANGES

Power Supply Coding	Power Supply Coding Changes
PS-915-1	LEAD DRESS CHANGED: Changed lead dress to prevent possibe shock hazard in event degausser coil shorts.
PS-915-4 and PS-919-2A	TO IMPROVE RELIABILITY OF Q-1R, Q-2G AND Q-3G: A diode, Part No. 48S191A08, added across R-13J (10 ohm).
PS-915-5 and PS-919-5	THERMAL DEGAUSSER THERMISTOR CHANGED TO NEW TYPE: Now consists of Part No. 31C66717A06 terminal strip and Part No. 6C65884A11 thermistor (NTC). Thermistor now located on terminal strip on top of power supply. Formerly located on degausser switch.

PS-915-6 and PS-919-6	TO IMPROVE HEAT DISSIPATION OF POWER RESISTORS, R-9J AND R-13J: Wattage dissipation of resistors increased to 20 watts (new Part No. 17S135868). Also, body of resistor clamped to chassis to further improve dissipation. Clamp Part No. 24C66393A04.
PS-915-6A and PS-919-6A	DEGUASSING SWITCH CHANGED TO A MORE RELIABLE TYPE: Degaussing switch, E-2J (Part No. 40D68091A03), changed to Part No. 40D68091A04. The new degaussing switch also included thermistor R-12J. Thermistor may be replaced separately if desired, Part No. 6C65884A11 (same as used in PS-915-5 and PS-919-5).

VIDEO IF PANEL "B" CODING CHANGES

Panel Coding	Panel Coding Changes
B-1	TO MINIMIZE 920KC BEAT IN PICTURE BY IMPROVING 4.5MC TRAP CIRCUIT AND AGC CIRCUIT: 4.5Mc trap changed to improved type, Part No. 24D68614A76. R-20B changed from 5.6K to 3.3K. R-39B changed from 750 to 470. R-7B changed from 2.2K to 1K. R-15B changed from 390 to 560. R-13B changed from 3.9K to 6.8K. Added lug for IF AGC feed back (lug Part No. 39S10184A09).
B-2	PANEL LEAD DRESS IMPROVED: Improved panel lead dress for better decoupling of 4.5Mc sound from +35V supply.
B-3	TO INCREASE RF AGC DELAY CONTROL RANGE: R-28B, RF delay control 3.5K changed to 25K, Part No. 18D66401A29. R-29B (8.2K) changed to 1K.
B-4	VIDEO IF HARMONIC SHIELDING ADDED TO TS-919 CHASSIS: Shield located on chassis between E panel and 3rd IF portion of B panel.
B-5	IF ETCHED PANEL LAYOUT CHANGED TO ACCEPT FTL PANELS: Added 1 pf, 10% FTL take-off capacitor, Part No. 21S115956.
B-6	MINIMIZE CHANNEL 6 AND 8 RF INTERFERENCE: L-11B, 7.5uh choke added in series with 4.5Mc sound output to (D) audio panel. Relocated FTI/FTL 35V B+ lug from outside IF shield to inside shield at junction of R-39B and R-21B (choke Part No. 24D66772A12).
B-7	TO IMPROVE 41.25MC TRAP TEMPERATURE DRIFT: Capacitor across 41.25Mc trap C-33 (75pf, 5%, 500V, NPO) changed to 75pf, 5%, 500V, N75 (Part No. 21S180D68).
B-7A only	(A) TO IMPROVE 39.75MC TRAP (L-3B) RANGE: Capacitor C-5B (22pf) changed to 18pf, 5%, NPO. (B) TO IMPROVE 1ST VIDEO AMPLIFIER, Q-1E, RELIABILITY: R-56B (330 ohm) added in series with Q-1E base.
B-7B and B-9A	TRANSISTOR REPLACEMENT: A small quantity of SE-5023 (Fairchild) transistors were used in the 1st and 2nd IF stages. If replacement is necessary, use Motorola type A1G-1.
B-8	IF PANEL LAYOUT REVISION: The IF panel layout was revised slightly to accept prior changes and different style of components.
B-8A	TO IMPROVE AGC TRACKING WITH LINE VOLTAGE CHANGES: Change R-24B (22K) to 18K, 5%, 1/2W. Remove R-30B, 10K, 10%, 1/2W. Add R-57B (3.6K, 5%, 1/2W) - see schematic diagram. Add R-58B (3.3K, 10%, 1/2W) - see schematic diagram. Add Zener diode E-3B, Part No. 48S137000 - see schematic diagram. Add R-59B (15K, 10%, 1/2W) - see schematic diagram.
B-8B	SAME AS B-7A (39.75MC TRAP ONLY).
B-9	DESIGN CHANGE: Delete B-7A (Part B). Change and increase working voltage of C-5B (8mf) from 15V to 50V, Part No. 23C66135A20.
B-9A	SAME AS B-7B.

AUDIO PANEL "D" CODING CHANGES

Panel Coding	Panel Coding Changes
D-1A and D-2	TO INCREASE AUDIO OUTPUT AND TONE CONTROL RANGE AND TO IMPROVE RELIABILITY BY REDUCING THE EFFECTS OF THE AUDIO DRIVER TRANSISTOR FROM DESTROYING THE AUDIO OUTPUT TRANSISTOR: The following changes were made: R-19D changed from 1meg, 5% to 560K, 5%. R-21D changed from 47K to 3.9K. R-22D changed from 56K, 5% to 4.7K, 5%. R-23D changed from 150K, 5% to 15K, 5%. R-24D changed from 2.2K to 1K. Removed R-18D (470K). Removed R-20D (470) and added a jumper in place of R-20D. C-20D changed from .22mf to .47mf, 100V. C-21D changed from .01mf to .047mf, 50V. Added R-18D (1K) connected in series with Q-3B base and C-19D (see schematic diagram).
D-3	PANEL RE-DESIGN: Re-designed panel to incorporate D-1A and D-2 changes.
D-3A and D-3B	RELIABILITY CHANGE: Audio amplifier transistor, Q-3D (4733) changed to A4B/7015, Part No. 48S137015. Audio driver transistor, Q-4D (4732) changed to A4A/7014, Part No. 48S137014.
D-3C	TRANSISTOR CHANGE: Audio amplifier, Q-3D (A4D) changed to A3K, Part No. 48S134997. Audio driver, Q-4D (A4A) changed to A3K, Part No. 48S134997. NOTE: A3K is recommended as replacement type for both Q-3D and Q-4D application.
D-4	SAME AS D-3A AND D-3B: Reduce working voltage of C-1D, C-6D, C-7D and C-23D from 100V to 50V, Part No. 21S180D84 (.1mf).

VIDEO AMPLIFIER PANEL "E" CODING CHANGES

Panel Coding	Panel Coding Changes
E-3	TO MINIMIZE PICTURE HIGHLIGHT LIMITING BY REDUCING VIDEO DETECTOR OUTPUT SIGNAL: R-25E changed from 1K to 1.5K. Also see E-4 coding.
E-4	TO INCREASE NOISE SEPARATOR SENSITIVITY TO ACCEPT THE REDUCED VIDEO DETECTOR OUTPUT SIGNAL (SEE E-3 CODING): The following changes were made: R-9E changed from 4.7K to 10K. R-18E changed from 39K to 33K. R-58E changed from 2.2K to 1.5K.
E-4A and E-5	TO MAKE BRIGHTNESS INSENSITIVE TO LINE VOLTAGE CHANGES: Q-7E, brightness stabilizer, changed from 4841 to 4953/A2K, Part No. 48S134953. R-4E changed from 6.8K to 27K. R-29E changed from 220 to 820. R-31E changed from 220, 5% to 68, 5%. R-52E changed from 27K, 5% to 12K, 5%.
E-6A and E-7	TO MAKE BRIGHTNESS STABLE OVER WIDER RANGE: The following changes were made: R-52E changed from 12K, 5% to 18K, 5%. R-31E changed from 68, 5% to 120, 5%. Transistor spacer and heat sink added to Q-7E, 4953/A2K.
E-7A	L-5E SUBJECT TO SHORTING DUE TO PHYSICAL SIZE ON TS-919 ONLY: L-5E removed and replaced with a miniature version, Part No. 24D67676A21.
E-8	DESIGN CHANGE: Suppression chokes L-6E and L-9E removed from circuit.

HORIZONTAL PANEL "F" CODING CHANGES

Panel Coding	Panel Coding Changes
F-1	TO REDUCE HORIZONTAL OSCILLATOR DRIFT WITH TEMPERATURE: R-9F (horizontal oscillator emitter) resistor, 150 ohm, 1/2 watt, 10% changed to 5% tolerance.
F-2	TO IMPROVE HORIZONTAL OUTPUT TRANSISTOR LIFE: Silicon grease added between heat sink and output transistors, Q-6F and Q-7F. Grease must be applied in an even coat to avoid hot spots on transistor case.
F-3	TO REDUCE BLANKER AMPLIFIER (Q-6L) BASE CURRENT: R-22F is changed from 100K, 1/2 watt, 10% to 470K, 1/2 watt, 10%.
F-3-1	TO IMPROVE HORIZONTAL BLANKING: R-22F (470K) in blanker amplifier (Q-6L) changed to 390K, 10%, 1/2 watt.
F-4	TO IMPROVE HORIZONTAL BLANKING: R-22F is changed from 470K, 10%, 1/2 watt to 330K, 10%, 1/2 watt.
F-3A and F-5	TO REDUCE POSSIBILITY OF HORIZONTAL OUTPUT TRANSISTOR THERMAL RUN-AWAY: A balancing coil (L-4F) added in series with bases of horizontal output pair to balance base current equally between pair (see schematic for alternate diagram).
F-6	TO REDUCE POSSIBILITY OF HORIZONTAL OSCILLATOR TRANSISTOR BREAKDOWN: Horizontal oscillator, Q-2F, 4842 changed to 7006/A3S, Part No. 48S137006.
F-7	TO REDUCE POSSIBILITY OF BALANCING COIL (L-4F) SHORTING TO HEAT SINK: Insulation tape added to heat sink.
F-8	TO CENTER HORIZONTAL OSCILLATOR TUNING RANGE: Resistor R-8F (4.7K) changed to 5.1K, 10%, 1/2 watt.

Panel Coding	Panel Coding Changes
F-9	DESIGN CHANGE: To improve reliability of horizontal output transistors; Arc gate transistor, Q-5F, removed from circuit. R-18F (68) changed to 47 ohm. R-17F (1.5K) changed to 390 ohm. Arc gate signal fed from junction of R-17F and R-16F to bottom of R-13F. See schematic diagram for wiring details.
F-10	ETCHED PANEL CHANGED: Etched panel changed to fiberglass-filled type.

PINCUSHION PANEL "G" CODING CHANGES

Panel Coding	Panel Coding Changes
G-1A only	TO IMPROVE RELIABILITY AND TO PROTECT Q-2G: Diode E-1G added between Q-3G collector and Q-2G base, Part No. 48S191A08.
G-1A, B and C	TO IMPROVE RELIABILITY AND TO PROTECT Q-2G: 10K resistor added between base and emitter of Q-2G.
G-1B	TO CENTER RANGE OF HIGH VOLTAGE AND HORIZONTAL SIZE CONTROL: R-8G changed from 8.2K to 10K.
G-2	PANEL LAYOUT CHANGED: Etched panel layout changed to incorporate G-1A, B and C changes.
G-3	DESIGN CHANGE: Delete diode added in G-1A, B and C change.
G-4	SAME AS G-1B.
G-4-2	CAPACITOR CHANGE: C-9G and C-5G changed to new type.

CONVERGENCE PANEL "H" CODING CHANGES

Panel Coding	Panel Coding Changes
H-2	TO PROVIDE A BETTER GROUND FOR PANEL: A ground strap was added from ground side of R-11H to vertical output transformer mounting screw.
H-3	TO INCREASE RELIABILITY AND PROVIDE A BETTER CONTACT FOR Q-1H (VERTICAL OUTPUT TRANSISTOR) EMITTER CURRENT: Panel layout was changed and a jumper added to by-pass terminal 1H. Plug, Part No. 39S10184A09, and wire jumper, Part No. 30V68618A60, added to reduce the possibility of vertical jitter due to poor connection at 1H.
H-4	TO IMPROVE RELIABILITY: Silicon diodes E-2 and E-3, Part No. 48S10062A02, changed to more reliable type, Part No. 48S191A08.
H-5	PANEL LAYOUT CHANGED: Panel layout changed to incorporate above changes.
H-6	PANEL CHANGES: Etched panel replaced with fiberglass-filled type.
H-6-2	CAPACITOR CHANGE: C-5H and C-3H changed to new type.
H-6-A	TO REDUCE VERTICAL JITTER: The following changes were made:
	R-9H (12) changed to 18 ohm, 10%, 1W.
	R-7H (2.2K) changed to 4.7K, 10%, 1/2W.

VIDEO DRIVE PANEL "L" CODING CHANGES

Panel Coding	Panel Coding Changes
L-5	TO ELIMINATE GREEN NOISE AND GREEN TRANSIENTS: A 22pf capacitor, C-21L, was added from base of green video driver, Q-3L, to ground.
L-6	TO REDUCE BRIGHTNESS AND RASTER FLUTTER: The following changes were made:
	C-11L changed from .047mf to .015mf.
	R-32L changed from 33K to 22K.
L-7	DESIGN CHANGE: Panel revised to accommodate PC (printed circuit) type capacitor, C-21. No change electrically.
L-7A only	DESIGN CHANGE: Diode, E-7L, added in series with base of Q-6L (M4842), blanking amplifier transistor (see schematic diagram).

VIDEO OUTPUT PANEL "M" CODING CHANGES

Panel Coding	Panel Coding Changes
M-1	FOR IMPROVED RELIABILITY OF VIDEO OUTPUT TRANSISTORS, Q-1M, Q-2M AND Q-3M: Type 4927/A1S changed to 7002/A3M, Part No. 48S137002.
M-1B and M-2	TO STABILIZE VIDEO OUTPUT TRANSISTORS BY BETTER HEAT DISSIPATION: Heat sink is changed to improved type, Part No. 26S66745A01.

FTI PANEL "P" CODING CHANGES

Panel Coding	Panel Coding Changes
P-2	PANEL CHANGED: Etched panel changed to improve component layout.
P-2A and P-3	TO IMPROVE FTI AMPLIFIER, DRIVER AND OUTPUT STABILITY: Circuit re-designed from solid state switch type to DC amplifier.
	R-1P changed from 100K to 27K.
	R-6P changed from 270 to 680.
	R-7P changed from 47K to 4.7K.
	R-9P changed from 1K to 3.3K.
	R-5P (470K) removed.
	R-5P (560) added from FTI output (Q-3P) base to center arm of FTI sensitivity control.

P-4	TO INCREASE FTI SENSITIVITY: The following changes were made:
	R-2P changed from 33K to 22K.
	R-5P and R-12P removed.
	Jumper added in place of R-12P (see schematic diagram).
P-5	PANEL DESIGN CHANGE: Panel revised to accommodate some of the above changes. No changes electrically.
P-6	TO REDUCE 45.75MC REGENERATION (APPEARS AS SMEAR ACROSS PICTURE - VISIBLE ONLY WHEN IMPROPERLY FINE TUNED) IN IF PANEL "B": Add two 7.5uh chokes, Part No. 24D66772A12, in series with FTI B+ lead and FTI output (reference numbers L-2 and L-3).

COLOR PANEL "S" CODING CHANGES

Panel Coding	Panel Coding Changes
S-1	TO IMPROVE ACC ACTION AND COLOR SYNC GATING: The following changes were made:
	R-19S changed from 3.3K to 4.7K.
	R-20S changed from 47K to 150K.
	Added L-6S coil in series with gating signal to emitter of Q-7S (L-6S - Part No. 24D67676A20).
S-2	DESIGN CHANGE TO DELETE COLOR INDICATOR LIGHT: The following changes were made:
	Removed Q-9S and R-45S.
	Relocated R-44S between Q-6S emitter and ground (see schematic diagram).
S-2A	TO REDUCE HUE DRIFT WITH TEMPERATURE: L-4S, color oscillator output collector coil, changed to improved type, Part No. 24D68771A04. Added C-43S, 390pf, 10% mica between C-57S and L-4S.
S-3	TO PREVENT COLOR KILLER FROM TURNING ON AT HIGH LINE VOLTAGE DUE TO TRANSISTOR BREAKDOWN: Q-4S changed from type 4733 to 4918/ A1L (Part No. 48S134918).
	NOTE: This does not include S-2 or S-2A change.
S-4	S-4 CHANGE INCORPORATES THE S-2, S-2A AND S-3 CHANGES.
S-5	PANEL CHANGE: New etched panel. No change electrically.
S-5A and S-6 (TS-915B-14 and TS-919B-12)	TO REDUCE HUE DRIFT: Q-11S changed from 4905/A1H to 4970/A2T, Part No. 48S134970. R-51S changed from 120 ohm to 220 ohm.
S-6	CAPACITOR CHANGE: C-20S and C-36S working voltage changed from 100V to 50V, Part No. 21S180D84 (.1mf). No change electrically.
S-6A	TO REDUCE HUE RANGE VARIATIONS: Color oscillator phase splitter circuit (Q-11S) re-designed.
	R-49 (27K, 10%, 1/2W) changed to 47K, 10%, 1/2W.
	R-52 (1.5K, 10%, 1/2W) changed to 2.7K, 10%, 1/2W.
	C-52 (82pf, 10%, N750) changed to 33pf, 10%, N750.

FTL PANEL "T" CODING CHANGES

Panel Coding	Panel Coding Changes
T-1	TO IMPROVE FTI (FINE TUNING INDICATOR) LIGHT SENSITIVITY: The following changes were made:
	C-4T changed from 82pf, 5% to 68pf, 5%.
	6.8 meg, 10%, 1/2 watt resistor added from collector to base of Q-4T.

T-2	PANEL LAYOUT CHANGED: Panel layout changed to incorporate T-1 change and improve parts lay out. Electrically the same as T-1.
T-2A and T-3	TO REDUCE UHF REGENERATION: Resistor R-18T, 100 ohm, 10%, 1/2 watt added in series with collector of FTL output Q-2T (A1Z).

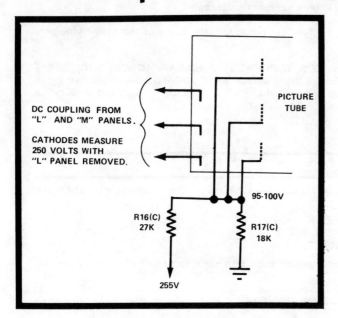

Fig. 4-5. Quasar TS-915-919 chassis CRT cathode circuits.

With one color out, or other colors predominant, look for a bad output transistor in video output panel M. If the circuitbreaker pops after 5 to 8 seconds and if there is no horizontal and, consequently, no raster, replace horizontal sweep panel F. If there is no raster, and you find no voltage at Terminal 3F in the primary of the flyback and pincushion panel G is good, then check H. REG Q1, and particularly R3(R), a 1K resistor that's somewhat buried and is often burned when Q1 turns red and "departs."

When high voltage is all right, but there is no video and no raster, check Q2E, the second video amplifier. If it is bad, replace with an exact substitute, since the ABL will not set up with a general part from someone's "suggested" parts substitution list. Use a PIC and nothing else.

In general panel troubles, watch for corrosion, loose contacts, and do remove colored shipping plugs when swapping plug-boards; otherwise, you might own several outright.

If there is more than normal cross modulation and adjacent-channel interference, try a B-25 or

later video IF panel, or change the tuner for one stocked after June 1970 (probably means a different RF amp).

New Nomax covered flybacks will prevent moisture absorption and give long life when replacements are necessary.

To remove a low-level buzz or hum in a TS-919, caused by ground loop currents, remove the ground chassis connection of C14D, a 0.33-mfd capacitor soldered to chassis, and resolder the foil end to the outer shield on the cable to the tone control. To gain access, remove the convergence panel.

Vertical jitter is often caused by a bad 1H contact on the H convergence panel. Cleaning and positive contacts will cure the problem, or you may install an H-3 panel or later version which will have a jumper wire.

If the raster remains at full maximum or is intermittent, it may or may not be due to a defective CRT. The DC supply circuit to the CRT grid could be the source of trouble. See Fig. 4-5. To check, remove video driver panel L. The CRT should go black. If the CRT remains at high brightness, measure or monitor the CRT grid 1 bias. In TS-915 chassis, measure at the white CRT lead on the terminal strip on the M panel assembly. In TS919-chassis, the terminal strip is conveniently located on the main chassis and beneath the fold-down back panel. If this voltage is around 95-100 volts—and stays there—the CRT may be the problem. But, if this voltage is high, then suspect the voltage divider R16(C) and R17(C). Suspect an intermittent open in R17.

Finally, a reminder to use a Variac when servicing solid-state chassis. A drop-in or drop-out at certain AC line potentials can tell many things, including overall operation of the regulation of the AC power supply.

TS-915-919 PANEL REPAIR TIPS

To assist the bench technician in locating the defective component, it is very important that outside technicians clearly identify, on the defective parts tag, the actual symptom observed prior to removing the panel from the color set.

SYMPTOM	REFERENCE NUMBER	DESCRIPTION	PART NUMBER
VIDEO IF PANEL "B"			
No video/audio may	C-37	Capacitor	23C66135A30
be distorted or weak	Q-3	Transistor	48S134937
video/audio may be	Q-4	Transistor	48S134910
distorted	Q-6	Transistor	48S134841
Weak or no horiz. sync	C-37	Capacitor	23C66135A30
No audio	Q-5	Transistor	48S134841

SYMPTOM	REFERENCE NUMBER	DESCRIPTION	PART NUMBER

AUDIO PANEL "D"

SYMPTOM	REFERENCE NUMBER	DESCRIPTION	PART NUMBER
No audio or distorted audio	C-18	Capacitor	23C66135A29
	Q-3	Transistor	48S134997
	Q-4	Transistor	48S134997
	R-26	Resistor	6G10260A83
	T-2	Ratio Det. Xfmr.	24D10282A02

VIDEO AMP PANEL "E"

SYMPTOM	REFERENCE NUMBER	DESCRIPTION	PART NUMBER
No brightness/no video	L-3	Delay Line	24V68614A49
	Q-1	Transistor	48S134841
	Q-2	Transistor	48S134910
	Q-5	Transistor	48S134933
AGC - picture overload on channel (snow between channel)	Q-5	Transistor	48S134933
No vertical sweep	Q-11	Transistor	48S134933

HORIZONTAL SWEEP PANEL "F"

SYMPTOM	REFERENCE NUMBER	DESCRIPTION	PART NUMBER
No H.V. - opens circuit breaker	C-20	Capacitor	8S10212D26
	E-1	Diode	48S134921
	E-2	Diode	48S134921
	Q-6	Transistor	48S134995
	Q-7	Transistor	48S134995

HORIZONTAL SWEEP PANEL "F" - CONTINUED

SYMPTOM	REFERENCE NUMBER	DESCRIPTION	PART NUMBER
No H.V./ckt. breaker does not open	Q-2	Transistor	48S137006
	Q-3	Transistor	48S134815
	Q-4	Transistor	48S134919
No horiz. lock	Q-1	Dual-diodes	48S134917

PINCUSHION PANEL "G"

SYMPTOM	REFERENCE NUMBER	DESCRIPTION	PART NUMBER
Narrow picture/ Can't adjust horiz. size or no high voltage	C-6	Capacitor	23C68466A02
	Q-2	Transistor	48S137041
	Q-3	Transistor	48S134918
	R-11	Resistor	6G10260A46
No vertical pincushion correction	Q-1	Transistor	48S134838

CONVERGENCE PANEL "H"

SYMPTOM	REFERENCE NUMBER	DESCRIPTION	PART NUMBER
Intermittent or no vertical sweep/ small vertical or vertical foldover	C-3	Capacitor	23C66135A15
	C-5	Capacitor	23C66135A01
	Q-2	Transistor	48S134842
	R-10	Resistor	17S10130A94
	T-1	Transformer	25D68538A01

VIDEO DRIVER PANEL "L"

SYMPTOM	REFERENCE NUMBER	DESCRIPTION	PART NUMBER
Weak or no red	Q-1	Transistor	48S134967
	T-1	Transformer	24D68517A10
Weak or no blue	Q-2	Transistor	48S134967
	T-2	Transformer	24D68517A10

Weak or no green	Q-3	Transistor	48S134967
	T-3	Transformer	24D68517A10
Low or no brightness	Q-5	Transistor	48S134953
	Q-4	Transistor	48S134953
Retrace lines (no blanking)	Q-6	Transistor	48S134842

VIDEO OUTPUT PANEL "M"

No red-weak red or excessive red	Q-1	Transistor	48S137002
	R-1	Resistor/ video driver	18D68764A01
No blue-weak blue or excessive blue	Q-2	Transistor	48S137002
	R-7	Resistor/ video drive	18D68764A01
No green-weak green or excessive green	Q-3	Transistor	48S137002
	R-12	Resistor/ video drive	18D68764A01

FTI PANEL "P"

FTI light stays on	L-1	Coil	24D68501A09
	L-2	Choke	24D66772A12
	Q-1	Transistor	48S134933
	Q-2	Transistor	48S134910
	Q-3	Transistor	48S134941
FTI light will not light	Q-1	Transistor	48S134933
	Q-2	Transistor	48S134910
	Q-3	Transistor	48S134941

COLOR PANEL "S"

No color or weak color	Q-1	Transistor	48S134841
	Q-2	Transistor	48S134841
	Q-7	Transistor	48S134841
	Q-10	Transistor	48S134842
	Q-12	Transistor	48S134841
Improper hue	Q-11	Transistor	48S134970

FTL PANEL "T"

FTI light stays on	C-5	Capacitor	21G10256A13
	L-3	Choke	24D66772A12
	Q-1	Transistor	48S134937
	Q-2	Transistor	48S134937
	Q-3	Transistor	48S137033

PANEL "T" COMPONENT LOCATION GUIDE

REF NO	LOC	REF NO	LOC
CAPACITORS		**TRANSISTORS**	
C-1T	A3	Q-1T	B3
C-2T	B3	Q-2T	B3
C-3T	B3	Q-3T	A2
C-4T	A3	Q-4T	A1
C-5T	B2		
C-6T	A2	**RESISTORS**	
C-7T	B2		
C-8T	B2	R-1T	A3
C-9T	B2	R-2T	B3
C-10T	B4	R-3T	A3
C-11T	A2	R-4T	B3
C-12T	A2	R-5T	B2
C-13T	A1	R-6T	A3
C-14T	A3	R-7T	B2
C-15T	A1	R-8T	A2
C-16T	A2	R-9T	A2
		R-10T	B2
MIS ELEC PARTS		R-11T	B2
		R-13T	B3
E-1T	A2	R-17T	A2
E-2T	B3		
		CONTROL	
COILS			
		R-12T	A1
L-1T	B1		
L-2T	A2	**TRANSFORMERS**	
L-3T	A1		
		T-1T	A3

PANEL "P" COMPONENT LOCATION GUIDE

REF NO	LOC
CAPACITORS	
C-1P	A1
C-3P	B1
C-4P	A1
C-5P	B2
MIS ELEC PARTS	
E-1P	B1
E-2P	B1
COIL	
L-1P	A1
TRANSISTORS	
Q-1P	B2
Q-2P	B2
Q-3P	A3
RESISTORS	
R-1P	A2
R-2P	A1
R-5P	A2
R-6P	B2
R-7P	A2
R-8P	A2
R-9P	B3
R-11P	A1
R-12P	B2
CONTROLS	
R-3P	B2

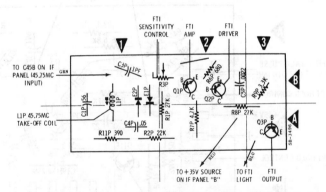

F.T.I. PANEL "P" - COMPONENT SIDE - CODED P-4

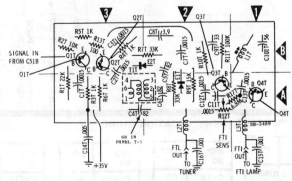

F.T.L. PANEL "T" - COMPONENT SIDE - CODED T-3

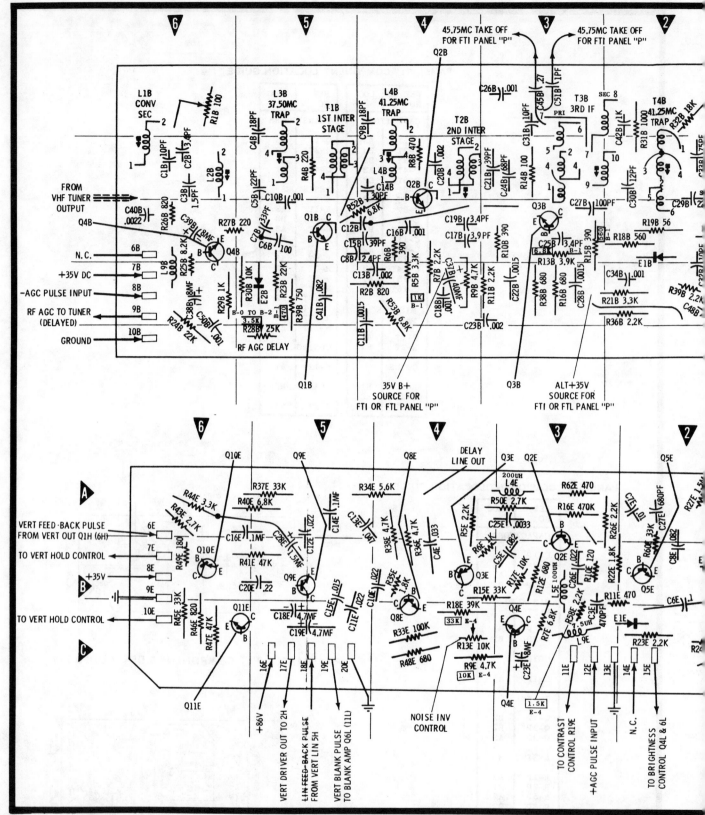

PANEL "B" COMPONENT LOCATION GUIDE

REF NO	LOC	REF NO	LOC	REF NO	LOC	REF NO	LOC	REF NO	LOC	REF NO	LOC	REF NO	LOC
CAPACITORS		C-7B	C5	C-14B	B4	C-21B	B3	C-28B	D3	C-35B	D1	C-42B	A2
C-1B	B6	C-8B	D5	C-15B	C4	C-22B	D3	C-29B	C2	C-36B	B1	C-43B	A1
C-2B	B6	C-9B	A5	C-16B	C4	C-23B	E3	C-30B	B2	C-37B	D4	C-44B	D2
C-3B	B6	C-10B	B5	C-17B	C4	C-25B	C3	C-31B	A3	C-38B	D6	C-45B	B3
C-4B	A5	C-11B	D4	C-18B	D4	C-26B	A3	C-32B	D2	C-39B	C6	C-46B	B1
C-5B	B5	C-12B	C5	C-19B	C4	C-27B	C3	C-33B	B2	C-40B	C6	C-47B	C1
C-6B	C5	C-13B	D4	C-20B	B4			C-34B	D2	C-41B	D5	C-48B	D2

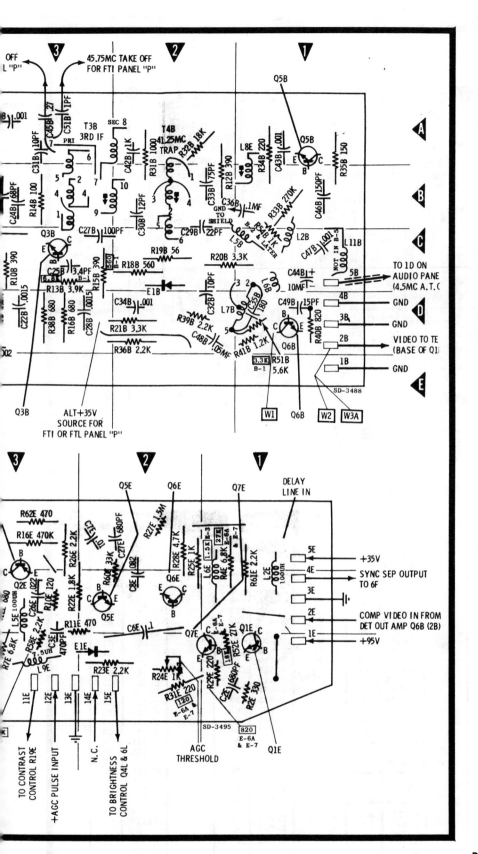

REF NO	LOC
C-49B	D1
C-50B	E6
MIS ELEC PARTS	
E-1B	D2
E-2B	D5
COILS	
L-1B	B6
L-2B	B6
L-3B	B5
L-4B	B4
L-5B	C1
L-6B	C1
L-7B	D2
L-8B	A1
L-9B	D6
L-10B	C1
TRANSISTORS	
Q-1B	C5
Q-2B	B4
Q-3B	C3
Q-4B	C6
Q-5B	B1
Q-6B	D1
CONTROLS	
R-1B	A6
R-28B	E5
RESISTORS	
R-2B	D4
R-4B	B5
R-5B	D4
R-6B	C4
R-7B	D4
R-8B	B4
R-9B	D4
R-10B	C3
R-11B	D3
R-12B	B2
R-13B	C3
R-14B	B3
R-15B	C3
R-16B	D3
R-18B	C2
R-19B	C2
R-20B	C2
R-21B	D2
R-23B	D5
R-24B	E6
R-25B	D6
R-26B	C6
R-27B	C6
R-29B	D6
R-30B	D5
R-31B	A2
R-32B	A2
R-33B	B1
A-34B	A1
R-35B	A1
R-36B	D2
R-37B	D2
R-38B	D3
R-39B	D5
R-40B	D1
R-41B	D2
R-51B	D1
R-52B	C4
R-53B	D4
TRANSFORMERS	
T-1B	A5
T-2B	A4
T-3BA	A3
T-3BB	A3
T-4B	A2

PANEL "E" COMPONENT LOCATION GUIDE

REF NO	LOC
CAPACITORS	
C-2E	C1
C-3E	C3
C-4E	B4
C-5E	B3

REF NO	LOC
C-6E	B2
C-7E	A2
C-8E	B2
C-10E	B4
C-11E	C5
C-12E	A5
C-13E	A4
C-14E	A5
C-15E	B5
C-16E	A6
C-18E	C5
C-19E	C5
C-20E	B5
C-23E	C3
C-25E	A3
C-26E	B3
C-27E	A2
C-28E	A5
MIS ELEC PARTS	
E-1E	C2
COILS	
L-2E	A1
L-4E	A3
L-5E	B3
L-6E	A1
L-9E	C3
TRANSISTORS	
Q-1E	B1
Q-2E	A3
Q-3E	B4
Q-4E	C3
Q-5E	B2
C-6E	B2
Q-7E	C2
Q-8E	C4
Q-9E	B5
Q-10E	B6
Q-11E	C6
CONTROLS	
R-13E	C4
R-24E	C2
RESISTORS	
R-2E	C1
R-4E	A1
R-5E	A4
R-6E	A4
R-7E	C3
R-9E	C3
R-10E	B3
R-11E	C3
R-12E	B3
R-15E	B3
R-16E	A3
R-17E	B3
R-18E	C4
R-22E	B3
R-23E	C2
R-25E	A2
R-26E	A3
R-27E	A2
R-28E	A2
R-29E	C1
R-31E	C2
R-33E	C4
R-34E	A4
R-35E	B4
R-36E	A4
R-37E	A5
R-38E	A4
R-40E	A5
R-41E	B5
R-43E	A6
R-44E	A6
R-45E	C6
R-46E	C6
R-47E	C6
R-48E	C4
R-49E	B6
R-50E	A3
R-52E	B1
R-56E	A3
R-58E	B3
R-60E	A2
R-61E	A1

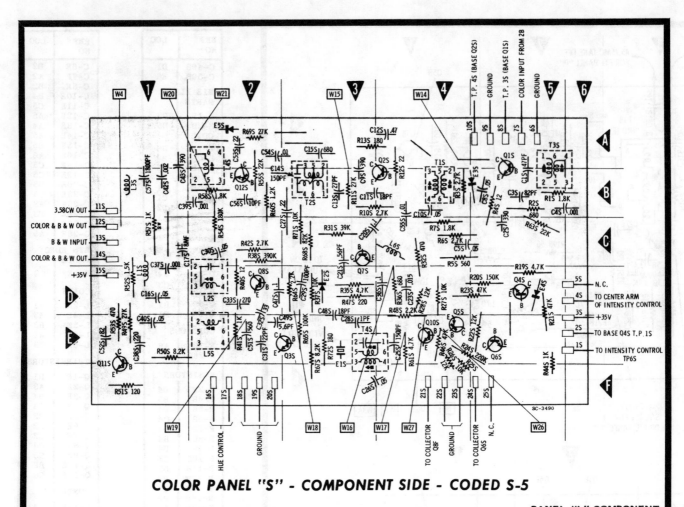

COLOR PANEL "S" - COMPONENT SIDE - CODED S-5

PANEL "L" COMPONENT LOCATION GUIDE

REF NO	LOC
CAPACITORS	
C-1S	B5
C-2S	C5
C-3S	B5
C-4S	B6
C-5S	C4
C-8S	B5
C-9S	B3
C-10S	B4
C-11S	B4
C-12S	A4
C-13S	B3
C-15S	A3
C-16S	D1
C-17S	C1
C-20S	C4
C-21S	C3
C-23S	D4
C-25S	E4
C-26S	F3
C-27S	B3
C-28S	E3
C-29S	D3
C-30S	C2
C-31S	E2
C-32S	D2
C-33S	D2
C-36S	D4
C-37S	D1
C-38S	E1
C-39S	B2
C-40S	E1
C-42S	B1
C-43S	A1
C-47S	D2
C-48S	E3
C-49S	E2

REF NO	LOC
C-51S	E2
C-52S	E1
C-53S	A2
C-54S	A2
C-55S	B4
C-56S	B2
C-57S	B1
MIS ELEC PARTS	
E-1S	E3
E-2S	D3
E-3S	B5
E-4S	D5
E-5S	A2
COILS	
L-1S	D1
L-2S	D2
L-3S	B1
L-4S	A2
L-5S	E2
L-6S	C4
TRANSISTORS	
Q-1S	A5
Q-2S	B4
Q-3S	E3
Q-4S	D5
Q-5S	E4
Q-6S	E5
Q-8S	C3
Q-10S	E4
Q-11S	F1
Q-12S	B2

PANEL "S" COMPONENT LOCATION GUIDE (CONT'D)

REF NO	LOC
RESISTORS	
R-1S	B5
R-2S	B5
R-3S	B5
R-4S	B5
R-5S	C4
R-6S	C4
R-7S	C4
R-10S	B4
R-11S	B3
R-12S	A4
R-13S	A4
R-19S	D5
R-20S	D5
R-21S	D5

REF NO	LOC
R-22S	E5
R-23S	D5
R-24S	E4
R-25S	E4
R-26S	F4
R-27S	D4
R-28S	D4
R-31S	C3
R-32S	C4
R-35S	D3
R-36S	D4
R-37S	D3
R-38S	C2
R-40S	D2

REF NO	LOC
R-41S	E2
R-42S	C2
R-44S	F5
R-46S	E4
R-47S	D3
R-48S	E4
R-49S	E1
R-50S	E1
R-51S	F1
R-52S	D1
R-53S	E1
R-54S	C2
R-55S	B2
R-57S	C1
R-58S	B2
R-60S	B2
R-61S	E4
R-64S	D3
R-63S	C5
R-65S	E3
R-67S	E3
R-68S	E3
R-69S	A2
R-71S	C3
R-72S	E3
TRANSFORMERS	
T-1S	B4
T-2S	B3
T-3S	A5
T-4S	E3

REF NO	LOC
CAPACITORS	
C-1L	A1
C-2L	A2
C-3L	B1
C-4L	A3
C-5L	C1
C-6L	C1
C-7L	B1
C-8L	B1
C-9L	C3
C-10L	E1
C-11L	E3
C-12L	E3
C-13L	D3
C-14L	C3
C-15L	D1
C-16L	A3
C-17L	D2
C-18L	B2
C-19L	B2
C-20L	D1
MIS ELEC PARTS	
E-1L	A1
E-2L	A1
E-3L	C1
E-4L	C1
E-5L	B1
E-6L	B1
COILS	
L-1L	C1

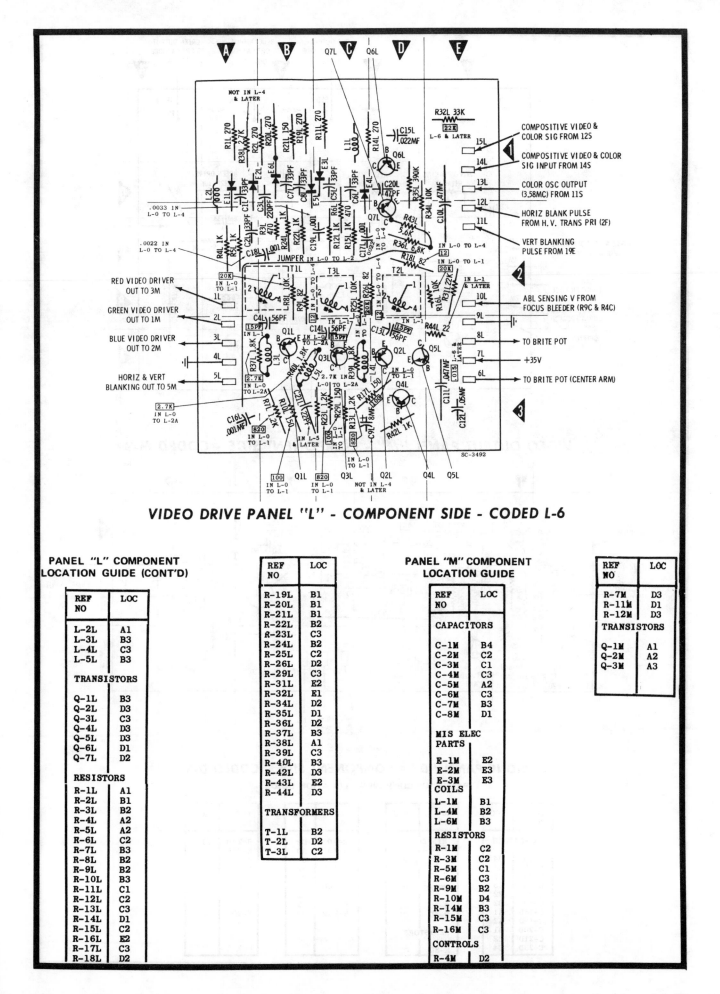

VIDEO DRIVE PANEL "L" - COMPONENT SIDE - CODED L-6

PANEL "L" COMPONENT LOCATION GUIDE (CONT'D)

REF NO	LOC
L-2L	A1
L-3L	B3
L-4L	C3
L-5L	B3
TRANSISTORS	
Q-1L	B3
Q-2L	D3
Q-3L	C3
Q-4L	D3
Q-5L	D3
Q-6L	D1
Q-7L	D2
RESISTORS	
R-1L	A1
R-2L	B1
R-3L	B2
R-4L	A2
R-5L	A2
R-6L	C2
R-7L	B3
R-8L	B2
R-9L	B2
R-10L	B3
R-11L	C1
R-12L	C2
R-13L	C3
R-14L	D1
R-15L	C2
R-16L	E2
R-17L	C3
R-18L	D2

REF NO	LOC
R-19L	B1
R-20L	B1
R-21L	B1
R-22L	B2
R-23L	C3
R-24L	B2
R-25L	C2
R-26L	D2
R-29L	C3
R-31L	E2
R-32L	E1
R-34L	D2
R-35L	D1
R-36L	D2
R-37L	B3
R-38L	A1
R-39L	C3
R-40L	B3
R-42L	D3
R-43L	E2
R-44L	D3
TRANSFORMERS	
T-1L	B2
T-2L	D2
T-3L	C2

PANEL "M" COMPONENT LOCATION GUIDE

REF NO	LOC
CAPACITORS	
C-1M	B4
C-2M	C2
C-3M	C1
C-4M	C3
C-5M	A2
C-6M	C3
C-7M	B3
C-8M	D1
MIS ELEC PARTS	
E-1M	E2
E-2M	E3
E-3M	E3
COILS	
L-1M	B1
L-4M	B2
L-6M	B3
RESISTORS	
R-1M	C2
R-3M	C2
R-5M	C1
R-6M	C3
R-9M	B2
R-10M	D4
R-14M	B3
R-15M	C3
R-16M	C3
CONTROLS	
R-4M	D2

REF NO	LOC
R-7M	D3
R-11M	D1
R-12M	D3
TRANSISTORS	
Q-1M	A1
Q-2M	A2
Q-3M	A3

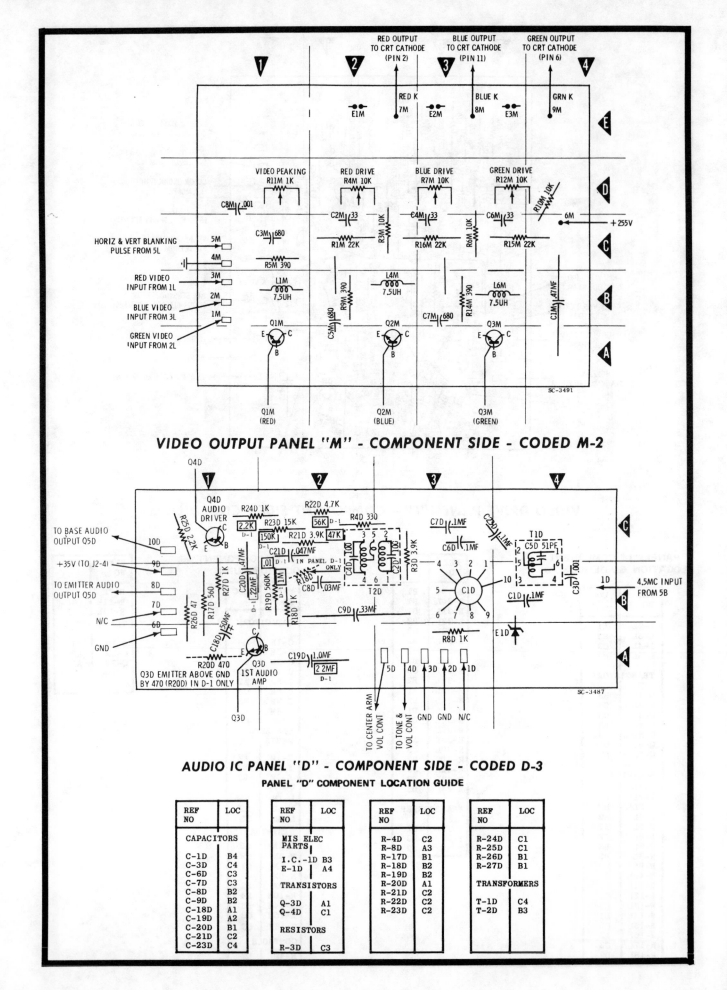

VIDEO OUTPUT PANEL "M" - COMPONENT SIDE - CODED M-2

AUDIO IC PANEL "D" - COMPONENT SIDE - CODED D-3

PANEL "D" COMPONENT LOCATION GUIDE

REF NO	LOC	REF NO	LOC	REF NO	LOC	REF NO	LOC
CAPACITORS		MIS ELEC PARTS		R-4D	C2	R-24D	C1
				R-8D	A3	R-25D	C1
C-1D	B4	I.C.-1D	B3	R-17D	B1	R-26D	B1
C-3D	C4	E-1D	A4	R-18D	B2	R-27D	B1
C-6D	C3			R-19D	B2		
C-7D	C3	TRANSFORMERS		R-20D	A1	TRANSFORMERS	
C-8D	B2			R-21D	C2		
C-9D	B2	TRANSISTORS		R-22D	C2	T-1D	C4
C-18D	A1			R-23D	C2	T-2D	B3
C-19D	A2	Q-3D	A1				
C-20D	B1	Q-4D	C1				
C-21D	C2						
C-23D	C4	RESISTORS					
		R-3D	C3				

REF NO	LOC
CAPACITORS	
C-1F	C2
C-2F	C2
C-3F	D2
C-4F	E3
C-5F	D3
C-6F	B3
C-7F	C3
C-8F	E3
C-9F	E1
C-10F	E2
C-11F	G1
C-12F	A3
C-13F	A2
C-16F	A2
C-17F	B2
C-18F	F1
C-19F	A1
C-20F	D4
C-21F	F3
C-22F	F3
C-23F	C2
MISC ELEC PARTS	
E-1F	A1
E-2F	B1
E-3F	F4
E-4F	F3
COILS	
L-1F	C4
L-2F	F3
L-3F	A1
TRANSISTORS	
Q-1F	C3
Q-2F	E2
Q-3F	F2
Q-4F	F2
Q-5F	G2
Q-8F	D1
CONTROL	
R-23F	F4
TRANSFORMER	
T-1F	D2
RESISTORS	
R-1F	D2
R-2F	C3
R-3F	C2
R-4F	C2
R-5F	D3
R-6F	D3
R-7F	D3
R-8F	D3
R-9F	D3
R-10F	E3
R-11F	F2
R-12F	F2
R-13F	E2
R-14F	E2
R-15F	G2
R-16F	F2
R-17F	F1
R-18F	F1
R-19F	B2
R-20F	B2
R-21F	D2
R-22F	B1
R-24F	E1
R-25F	F3
R-26F	C1
R-27F	D1

HORIZONTAL SWEEP PANEL "F" - COMPONENT SIDE - CODED F-6

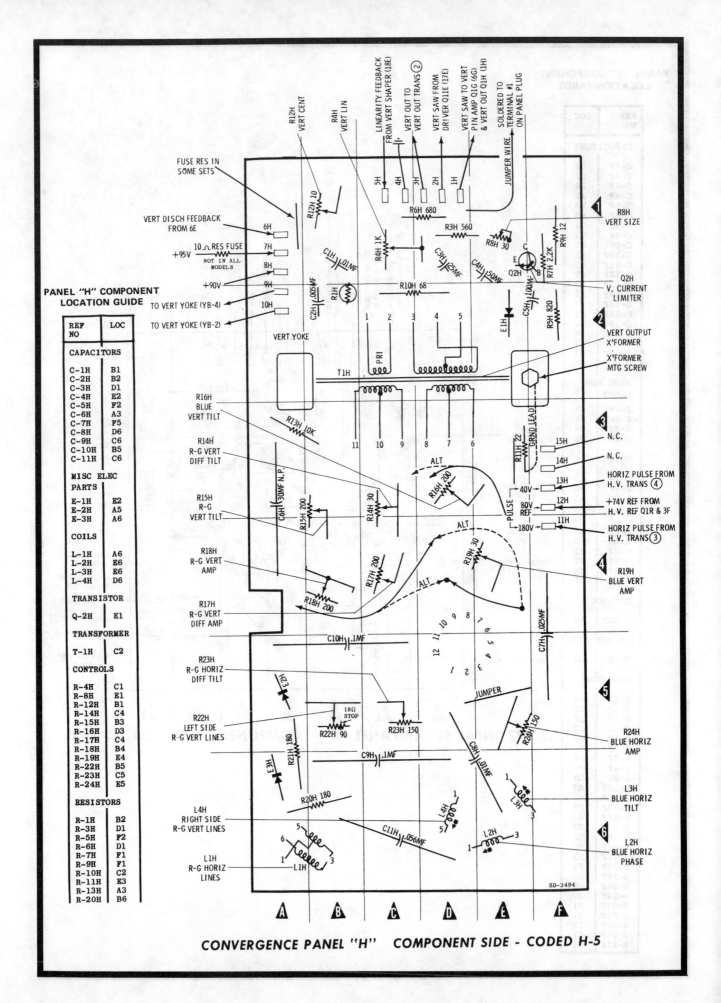

CONVERGENCE PANEL "H" COMPONENT SIDE - CODED H-5

PANEL "H" COMPONENT LOCATION GUIDE

REF NO	LOC
CAPACITORS	
C-1H	B1
C-2H	B2
C-3H	D1
C-4H	E2
C-5H	F2
C-6H	A3
C-7H	F5
C-8H	D6
C-9H	C6
C-10H	B5
C-11H	C6
MISC ELEC PARTS	
E-1H	E2
E-2H	A5
E-3H	A6
COILS	
L-1H	A6
L-2H	E6
L-3H	E6
L-4H	D6
TRANSISTOR	
Q-2H	E1
TRANSFORMER	
T-1H	C2
CONTROLS	
R-4H	C1
R-8H	E1
R-12H	B1
R-14H	C4
R-15H	B3
R-16H	D3
R-17H	C4
R-18H	B4
R-19H	E4
R-22H	B5
R-23H	C5
R-24H	E5
RESISTORS	
R-1H	B2
R-3H	D1
R-5H	F2
R-6H	D1
R-7H	F1
R-9H	F1
R-10H	C2
R-11H	E3
R-13H	A3
R-20H	B6

SD-3494

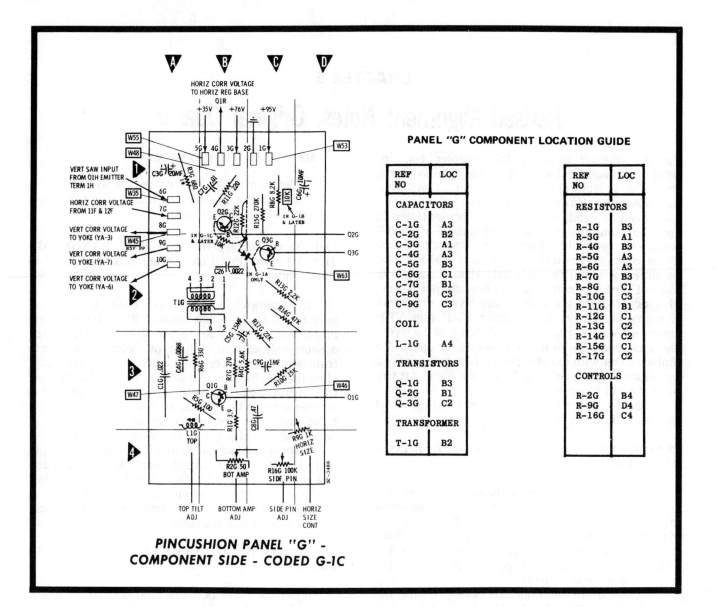

PANEL "G" COMPONENT LOCATION GUIDE

REF NO	LOC
CAPACITORS	
C-1G	A3
C-2G	B2
C-3G	A1
C-4G	A3
C-5G	B3
C-6G	C1
C-7G	B1
C-8G	C3
C-9G	C3
COIL	
L-1G	A4
TRANSISTORS	
Q-1G	B3
Q-2G	B1
Q-3G	C2
TRANSFORMER	
T-1G	B2

REF NO	LOC
RESISTORS	
R-1G	B3
R-3G	A1
R-4G	B3
R-5G	A3
R-6G	A3
R-7G	B3
R-8G	C1
R-10G	C3
R-11G	B1
R-12G	C1
R-13G	C2
R-14G	C2
R-15G	C1
R-17G	C2
CONTROLS	
R-2G	B4
R-9G	D4
R-16G	C4

**PINCUSHION PANEL "G" -
COMPONENT SIDE - CODED G-1C**

CHAPTER 5

Revised Alignment Notes: Original Quasar

For the original Quasars, this chapter contains setup and alignment data that was not available when Vol. 1, Motorola Color TV Manual (TAB 509), was published. However, since most of the basics have been covered, we confine our presentation to new information and to combination radio-TV receivers. You will find some stereo service notes, stylus replacement information, and AM-FM and amplifier schematics included, should repair of some of the large combinations become necessary. Also included are schematics and layout diagrams of many of the original Quasar UHF and VHF tuners, in which transistors will have to be eventually changed due to leakage, shorts, or opens. Do replace them with exact types or factory-approved equivalents from a reliable distributor's stock if you want the old tuner to live again.

Figs. 5-1 through 5-4 will be in locating panel and control locations.

COLOR SYNC, OSCILLATOR & HUE RANGE ADJUSTMENTS

1. **Color sync filter transformer** (T-4S, located on color panel S). Tune to a color telecast. Connect a meter across ACC rectifier diode E-2S. Adjust the sync filter bottom core three-sixteenth inch from the bottom of the can. Adjust the sync filter top (T4S) core for a maximum positive reading on the meter. Remove the meter.

2. **Color oscillator tank coil** (L5S). Ground the pulse limiter base Terminal 21S. Oscillator is now free-running. Place a short collector to emitter on color killer output transistor Terminals 24S and 1S. This supplies turn-on bias for the second color IF amplifier. Adjust color oscillator tank coil L5S until it is on frequency. This is indicated when the color and black-and-white are superimposed. Remove all jumpers.

3. **Color oscillator output coil** (L4S, hue range adjustment). Set the hue and tint controls to the mid-mechanical position. Adjust the color oscillator output coil (L4S) for the correct fleshtones.

Alternate Method

1. **Color sync filter transformer** (T4S). Tune the receiver to a color telecast. Adjust the sync filter (T4S) top core for minimum color as viewed on the screen.

2. **Color oscillator tank coil** (L5S). Adjust color oscillator tank coil L5S for best sync.

3. **Color oscillator output coil** (L4S, hue range adjustment). Set the hue and tint controls to the mid-mechanical position. Adjust the oscillator output coil (L4S) for the correct fleshtones.

4.5-MHz TRAP ADJUSTMENT

1. Tune correctly to a color telecast.
2. Adjust the 4.5-MHz trap, L7B, for minimum 920-kHz beat in the picture.

COLOR OSCILLATOR AND HUE RANGE ADJUSTMENTS (For Color Panel S, Coded S-8 And Later Only)

1. **Oscillator adjustment.** Tune to a color telecast. Ground pulse limiter base (Q10S) Terminal 21S; the oscillator is now free-running. Place a short collector to emitter on color killer output transistor (Q5S), Terminals 24S and 1S. This supplies turn-on bias for the second color IF amplifier Q2S. Adjust color oscillator tank coil L2S until it is on frequency. This is indicated when the color and black-and-white are superimposed. (Tune slowly—adjustment is sharp.) Remove all jumpers.

2. **Hue range adjustment.** Color oscillator output coil (L4S hue range adjustment). Set the hue and tint controls to the mid-mechanical position. Adjust the color oscillator output coil (L4S) for correct fleshtones. Note: If a color-bar generator is used, tune for third bar maximum red.

3.58-MHz TRAP ADJUSTMENT (Applicable To All Panels)

1. Set the tuner between channels (snow free).
2. Turn off the green and blue G2 controls. Adjust T1L for minimum 3.58 beat on a red field.
3. Turn off the red G2 and turn up the blue G2 to produce a blue field. Adjust T2L for minimum 3.58 beat on a blue field.
4. Turn off the blue G2 and turn on the green G2 for a green field. Adjust T3L for minimum 3.58 beat on the green field.
5. Readjust all G2 controls for proper black-and-white tracking.

SOUND ALIGNMENT PROCEDURE (Applicable To Audio Panels Coded D-5 And Later)

Reduce the signal input into the receiver by disconnecting one side or both antenna leads from the receiver. The signal should be reduced considerably until some background noise is present.

1. Adjust the primary coil (bottom core nearest panel) of T2D for maximum audio.

The following foldout section consists of pages 97 through 132. ➔

STEP	SWEEP GENERATOR	MARKER GENERATOR	ADJUST	REMARKS
1.	Set sweep generator to 44Mc. Adjust sweep width to desired response. Connect generator high side through a series connected .002 mf capacitor and 47 ohm resistor to TP-3B, 2nd IF collector low side to chassis. Use just enough output to obtain .4V PP useable pattern on scope.	Connect marker high side to TP-3B, 2nd IF collector through a series connected .002 capacitor and 47 ohm resistor low side to chassis. Set marker generator to near maximum for this step only.	Clamp the IF AGC to +5V DC. T-4B, 41.25Mc trap - L-8B 41.25Mc phase coil	Expand scope pattern to expose 41.25Mc trap (see curve 1, Figure 1). Adjust T-4B for frequency and L-8B for attenuation. When T-4B and L-8B are properly set, marker vanishes.

3RD IF ALIGNMENT

STEP	SWEEP GENERATOR	MARKER GENERATOR	ADJUST	REMARKS
2.	Set sweep generator to 44Mc. Adjust sweep width to desired response. Connect generator high side to TP-3B, 2nd IF collector through a series connected .002mf capacitor and 47 ohm resistor. Low side to chassis. Use just enough output to obtain a 2V PP pattern on scope.	Radiate marker signals into circuit by wrapping unshielded marker leads around unshielded sweep leads.	T-3B, 3rd IF primary and secondary.	Adjust T-3B primary and secondary to place 42.67Mc marker nearest the most negative portion of response curve (see curve 2, Figure 1) since there is inter-action between 41.25 Mc traps and 3rd IF. Repeat Steps 1 and 2 until the markers and response appears as in curve 3, Figure 1.

ADJACENT CHANNEL TRAPS

STEP	SWEEP GENERATOR	MARKER GENERATOR	ADJUST	REMARKS
3.	Not used.	Connect marker generator high side to mixer base T.P. M through .002mf capacitor and 10 ohm resistor. Low side to chassis.		IMPORTANT: Adjust varible DC bias supply to clamp IF AGC at precisely +3.2V DC.
		IMPORTANT: Use just enough generator output to indicate trap location.	L-2B and rheostat R-1B, 47.25Mc trap.	Set marker to 47.25Mc. Adjust L-2B and R-1B for maximum positive voltage reading on VTVM at test point 2B (repeat several times).
			L-3B, 39.75Mc trap.	Set marker to 39.75Mc. Adjust L-3B for maximum positive reading on VTVM.
			L-4B 35.25Mc trap.	Set marker to 35.25Mc. Adjust L-4B for maximum positive reading on VTVM.
4.	Sweep generator set to 44Mc. Adjust sweep width for desired response. Connect generator high side through a series connected .002mf capacitor and 47 ohm resistor to mixer base T.P. M . Low side to chassis. Use just enough generator output to obtain a 2V PP pattern on scope.	Radiate marker generator into circuit by wrapping unshielded marker leads around unshielded sweep leads.	L-1B convertor secondary.	Set marker generator to 45.75Mc. Adjust L-1B to place 45.75 Mc marker near 60% (see curve 4, Figure 1).
			T-2B, 2nd IF interstage transformer.	Set marker generator to 42.67Mc. Adjust T-2B to place 42.67 Mc marker near 10% (see curve 4, Figure 1).
			Retouch L-2B and rheostat, R-1B, 47.25Mc trap.	Set marker generator to 47.25Mc. Expand scope pattern to expose 47.25Mc trap. Retouch L-2B, 47.25Mc trap, and rheostat, R-1B, to null out marker. Re-set scope for 2V PP pattern. Check response curve for proper shape and marker placement.

Chart 5-1. IF alignment procedure.

STEP	SWEEP GENERATOR	MARKER GENERATOR	ADJUST	REMARKS
1.	Set sweep generator to 3Mc. Connect RCA WG-295C multi-marker in series with sweep output (see Figure 2). Connect a 10 ohm, 1/2 watt resistor across the output leads of the multi-marker. Connect high side of multi-marker through a 10 mfd capacitor to base of 2nd color IF terminal ⑩S. Low side to chassis. Adjudt sweep width for desired response.	Not used.	T-2S, 2nd color IF transformer (identified on panel shield).	Adjust variable DC bias supply connected to IF AGC TP-2B to +5V DC. Adjust T-2S top and bottom cores for proper marker placement and response as shown in curve 5. Adjust top and bottom cores so they are furthest from each other while still maintaining a symmetrical response. Keep generator output low to prevent distortion from overload.
		1ST AND 2ND COLOR IF		
2.	Set sweep generator to 3Mc. Connect RCA WG-295C multi-marker in series with sweep output. Connect a 10 ohm, 1/2 watt resistor across the output leads of multi-marker. Connect high side of multi-marker through a 10 mfd capacitor to base of 1st color IF terminal ⑧S.	Not used.	T-1S, 1st IF transformer (identified on shield).	Adjust T-1S core for desired response curve and marker placement as shown in curve 6. Adjust core on peak that positions core closest to the panel. Keep generator output low to prevent distortion from overload.
		OVERALL COLOR IF		
3.	Remove 10 ohm resistor across the multimarker output leads. Set generator to 3Mc. Connect RCA WG-295C multimarker in series with sweep output. Connect multimarker output to RCA WG-304B (RF modulator - see Figure 2). Connect output of RF modulator through .002 capacitor and 47 ohm resistor to mixer base T.P. Ⓜ on VHF tuner (see Figure 1).	Set marker generator to 45.75Mc. Connect output of marker generator to RCA WG-304B (RF modulator - see Figure 2).	T-3S, color take-off (CTO) transformer (identified on shield).	Adjust variable DC bias supply (connected to TP-2B) to precisely +3.2V DC (see Figure 1). Adjust T-3S to place markers and response as shown in curve 1. T-3S top and bottom cores should be adjusted furthest from each other while maintaining desired response. Both cores affect markers.

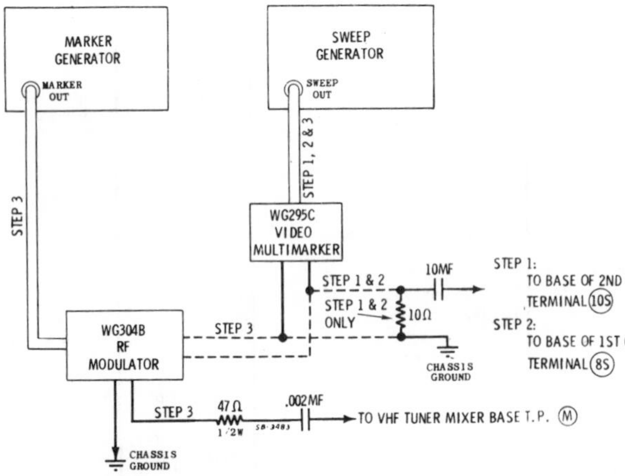

Chart 5-2. Color IF alignment procedure.

2. Adjust the secondary coil (top core) of T2D for the best sound with the least noise.

3. Repeat Steps 1 and 2 to optimize as necessary.

AFC ADJUSTMENT

Check the operation of the AFC as follows: Defeat the AFC by pulling out on the VHF fine-tuning knob. Fine tune to a station for the best color picture (just off burble). Push in the fine-tune knob to activate the AFC circuit. The picture should remain the same without severe detuning. If a loss of color or detuning is present, the following adjustments are recommended.

1. Pull out the AFC knob and adjust the fine tuning for the best color picture.

2. Adjust a VTVM for zero center scale.

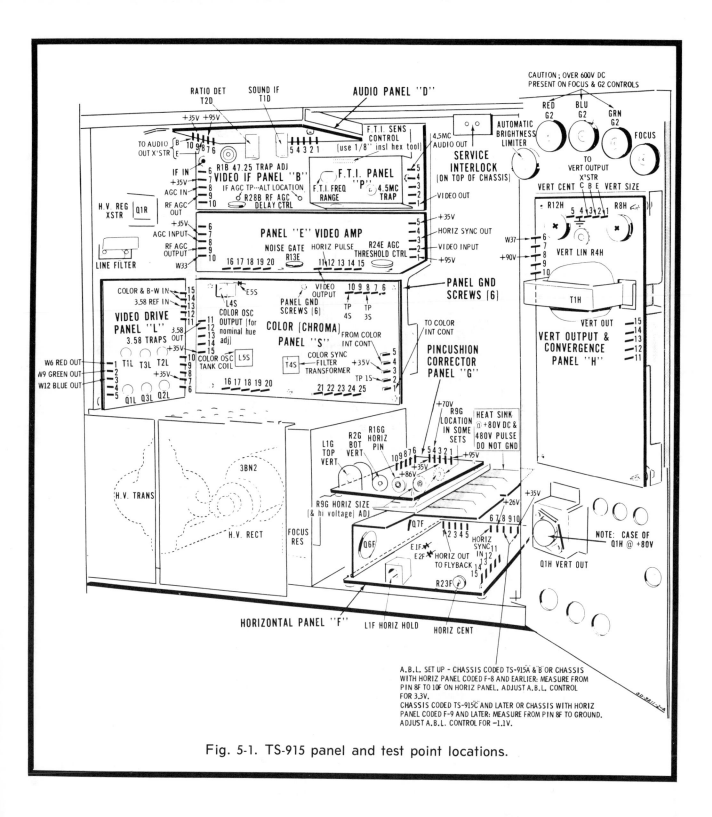

Fig. 5-1. TS-915 panel and test point locations.

TS-919: Connect the VTVM to the center conductor of the shielded cable attached to the rear of the UHF tuner (AFC lead).

TS915: Connect the VTVM across the AFC switch.

3. Push in the AFC knob to activate the AFC.

4. Adjust the top core of discriminator transformer T1 to the top of the can.

5. Adjust the bottom core to the bottom of the can.

6. Adjust the bottom core to the first peak reading on the VTVM.

7. Adjust the top core for zero center scale on VTVM.

8. Disconnect the VTVM and observe the AFC action by pulling out and pushing in AFC switch. If tuned properly, little or no detuning will be noticed

135

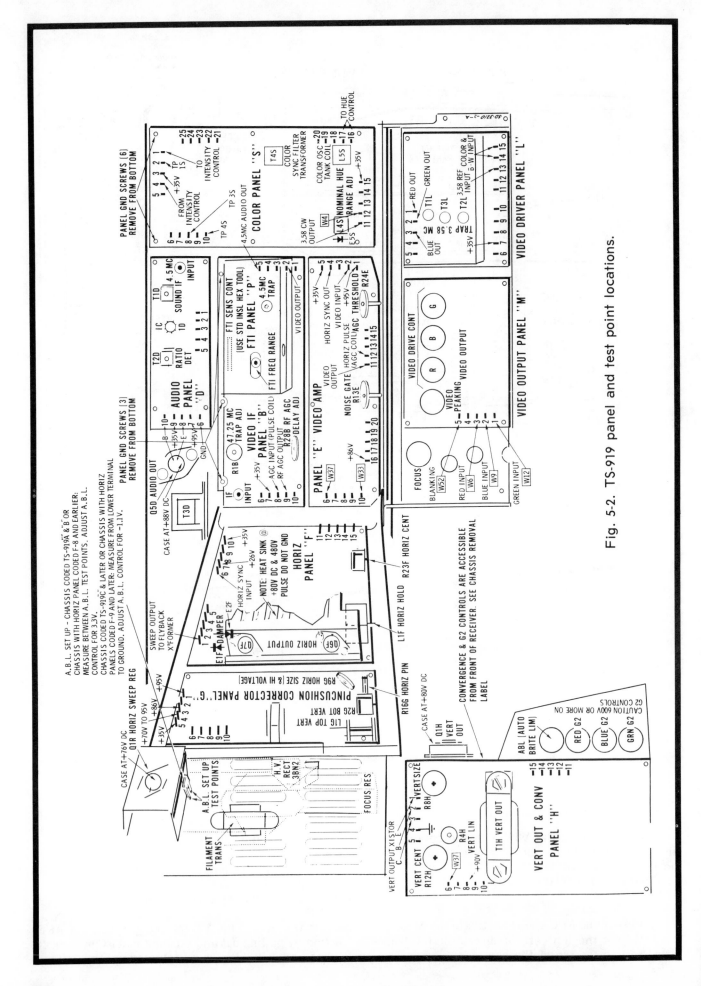

Fig. 5-2. TS-919 panel and test point locations.

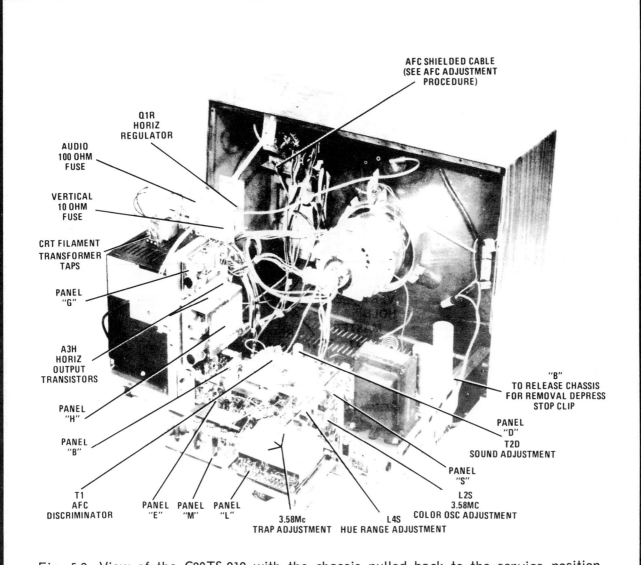

AFC SHIELDED CABLE
(SEE AFC ADJUSTMENT
PROCEDURE)

Q1R
HORIZ
REGULATOR

AUDIO
100 OHM
FUSE

VERTICAL
10 OHM
FUSE

CRT FILAMENT
TRANSFORMER
TAPS

PANEL
"G"

A3H
HORIZ
OUTPUT
TRANSISTORS

PANEL
"H"

PANEL
"B"

T1
AFC
DISCRIMINATOR

PANEL
"E"

PANEL
"M"

PANEL
"L"

3.58Mc
TRAP ADJUSTMENT

L4S
HUE RANGE ADJUSTMENT

L2S
3.58MC
COLOR OSC ADJUSTMENT

PANEL
"S"

PANEL
"D"
T2D
SOUND ADJUSTMENT

"B"
TO RELEASE CHASSIS
FOR REMOVAL DEPRESS
STOP CLIP

Fig. 5-3. View of the C20TS-919 with the chassis pulled back to the service position.

on the picture. If detuning is severe, repeat Steps 6, 7 and 8.

Schematic and layout diagrams for the tuners used in TS-915-919 chassis appear in Figs. 5-8 through 5-15.

STEREO PICKUP SERVICE NOTES

Stylus Replacement

To remove the stylus, open the tone arm restriction clip (see Fig. 5-8). Lift the tone arm to a vertical position and rotate the stylus selector to the 78 RPM position. Grasp stylus with the fingers and with a gentle motion, twist it up so the tip end raises and the rear portion snaps out from under the retaining clip.

To replace the stylus, check tone arm restriction clip. It should be in the open position;

then lift the tone arm to a vertical position. Holding the new stylus in position with the fingers, snap the rear portion of the stylus under the retaining clip. Be sure the stylus rests in the V of the cartridge; then return the tone arm to the normal position and place the restriction clip in the closed position. A replacement stylus is available under the following part number: 92-812DS.

Cartridge Replacement

The cartridge used in these models is a plug-in type. To remove it, grasp the cartridge housing and carefully pull out from the tone arm. The cartridge is not removable from its housing.

An inter-unit wiring diagram appears in Fig. 5-17. Stereo amplifier and tuner schematics are presented in Figs. 5-18 through 5-21.

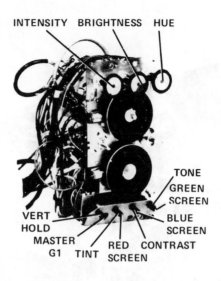

INTENSITY   BRIGHTNESS   HUE

TONE
GREEN
SCREEN

BLUE
SCREEN

VERT
HOLD
MASTER   RED   CONTRAST
G1   TINT   SCREEN

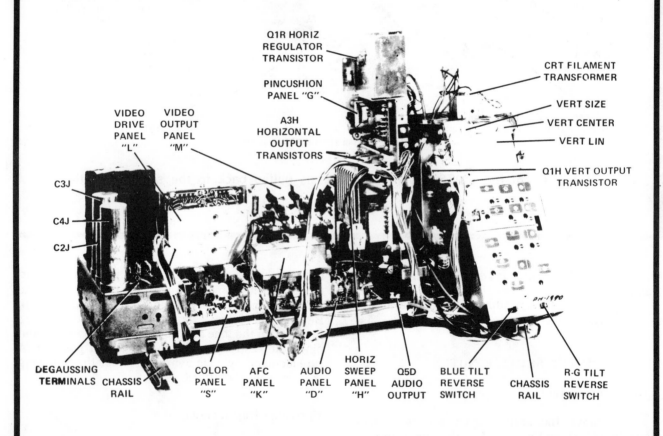

Q1R HORIZ
REGULATOR
TRANSISTOR

PINCUSHION
PANEL "G"

A3H
HORIZONTAL
OUTPUT
TRANSISTORS

VIDEO   VIDEO
DRIVE   OUTPUT
PANEL   PANEL
"L"   "M"

C3J

C4J

C2J

CRT FILAMENT
TRANSFORMER

VERT SIZE

VERT CENTER

VERT LIN

Q1H VERT OUTPUT
TRANSISTOR

DEGAUSSING   COLOR   AFC   AUDIO   HORIZ   Q5D   BLUE TILT   R-G TILT
TERMINALS CHASSIS PANEL   PANEL   PANEL   SWEEP   AUDIO   REVERSE   CHASSIS   REVERSE
           RAIL   "S"   "K"   "D"   PANEL   OUTPUT   SWITCH   RAIL   SWITCH
                                            "H"

Fig. 5-4. Panel and control locations, front view, C20TS-919.

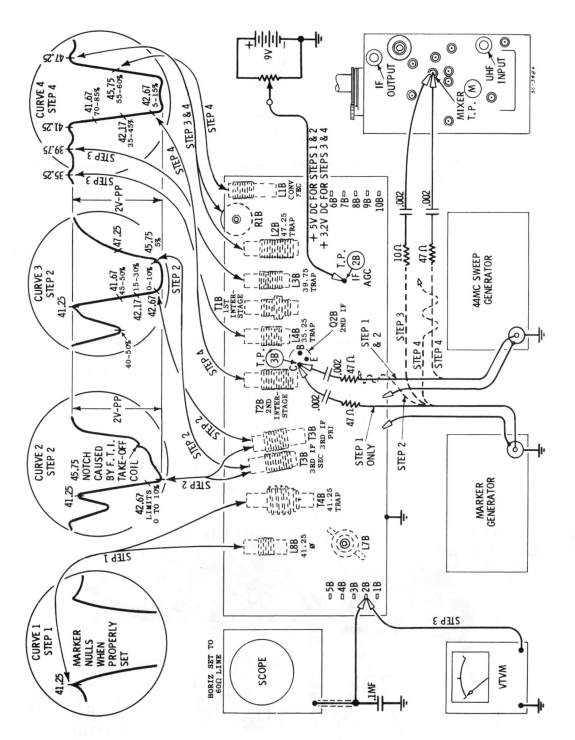

Fig. 5-5. Generator and scope connections, video IF alignment.

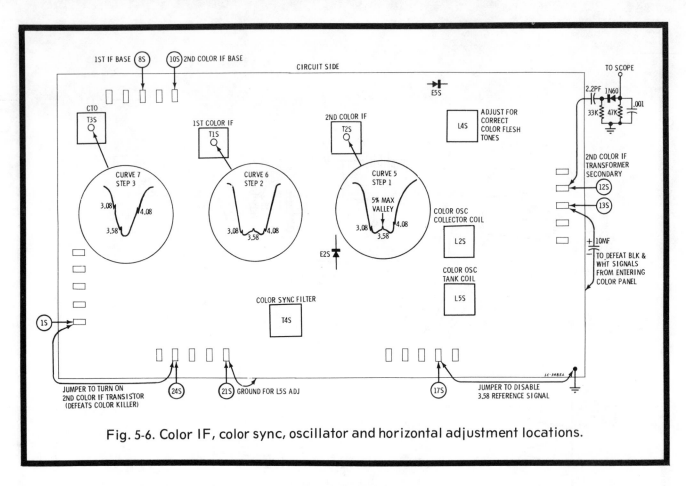

Fig. 5-6. Color IF, color sync, oscillator and horizontal adjustment locations.

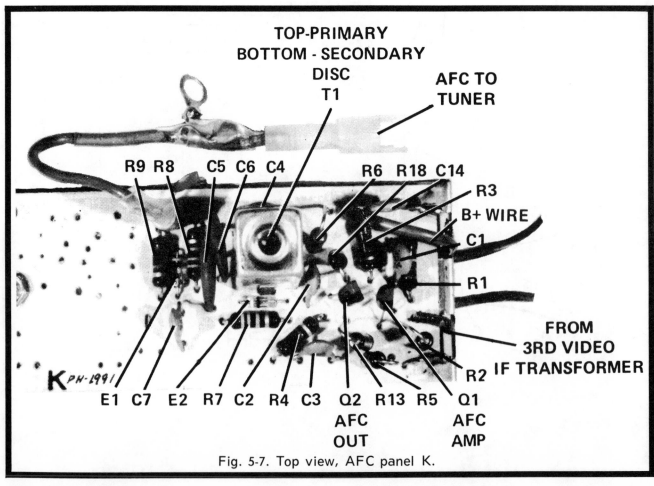

Fig. 5-7. Top view, AFC panel K.

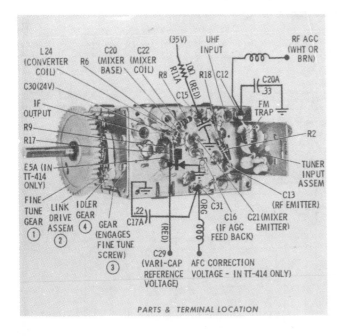

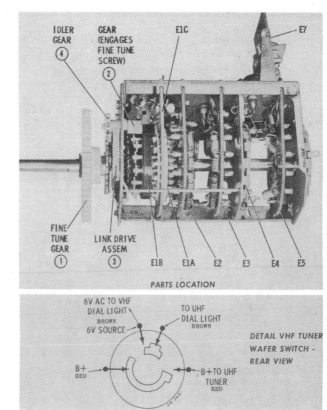

Fig. 5-8. Top and side views of VHF tuners TT-402 and TT-414.

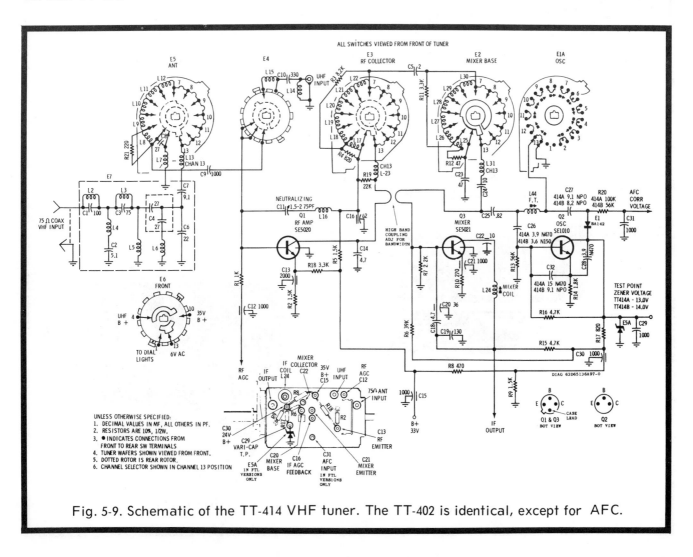

Fig. 5-9. Schematic of the TT-414 VHF tuner. The TT-402 is identical, except for AFC.

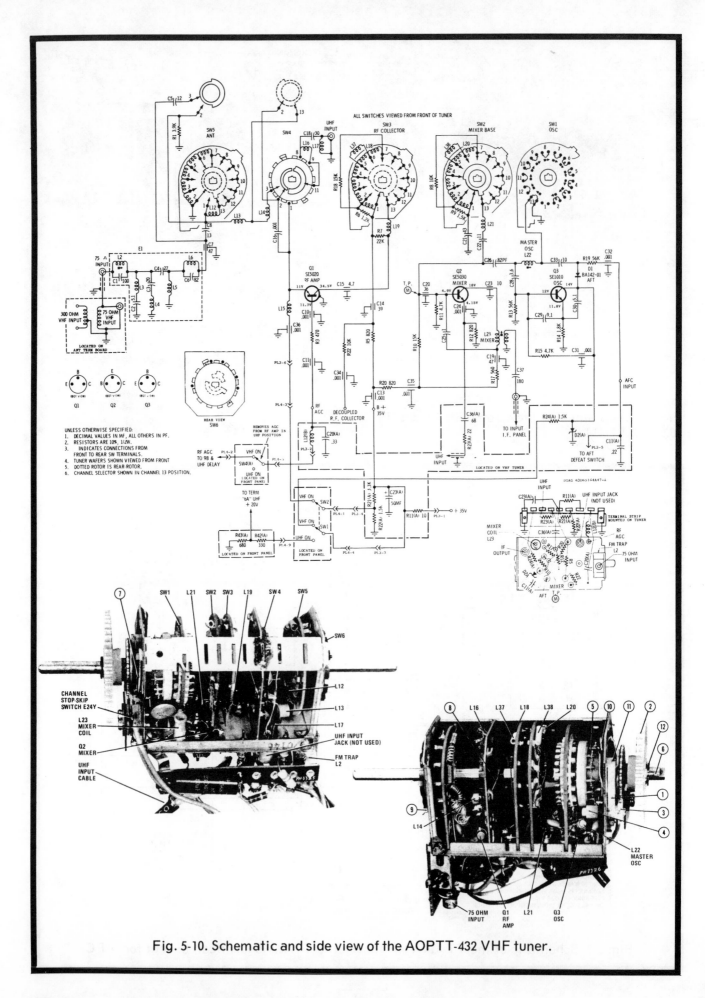

Fig. 5-10. Schematic and side view of the AOPTT-432 VHF tuner.

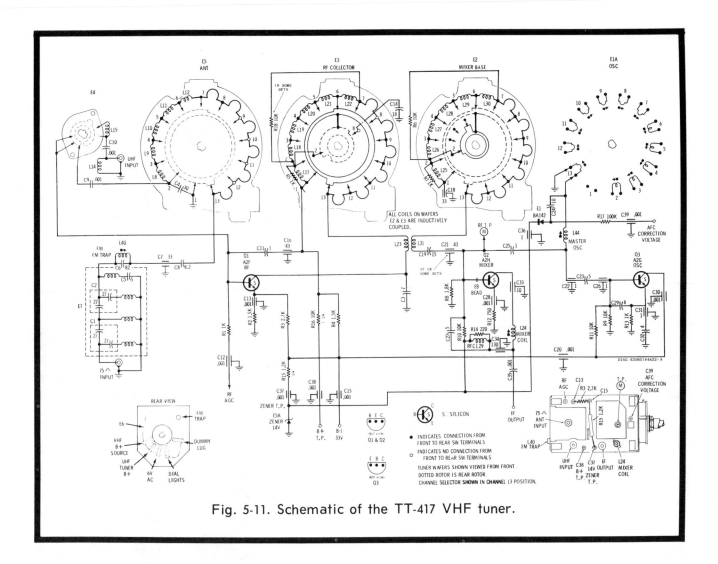

Fig. 5-11. Schematic of the TT-417 VHF tuner.

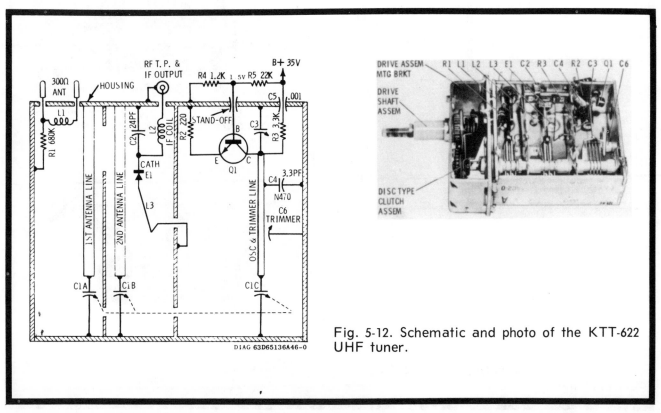

Fig. 5-12. Schematic and photo of the KTT-622 UHF tuner.

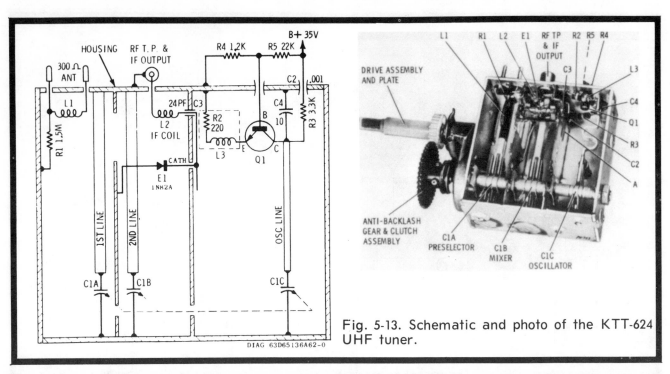

Fig. 5-13. Schematic and photo of the KTT-624 UHF tuner.

DIAG 63D65136A62-0

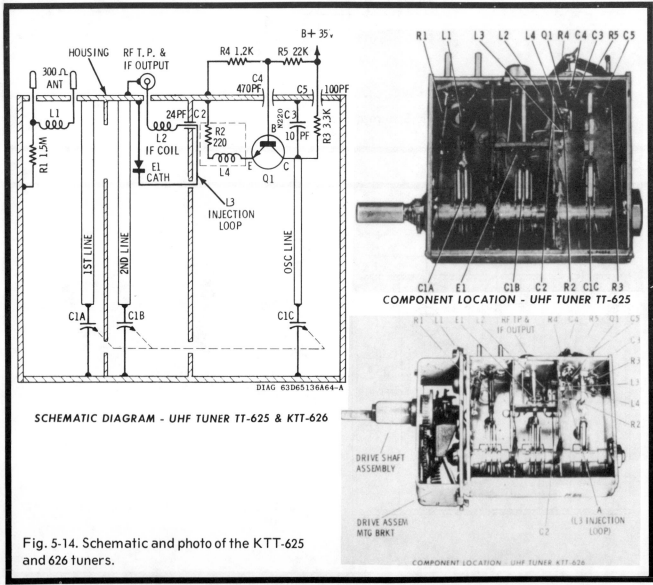

COMPONENT LOCATION - UHF TUNER TT-625

SCHEMATIC DIAGRAM - UHF TUNER TT-625 & KTT-626

DIAG 63D65136A64-A

Fig. 5-14. Schematic and photo of the KTT-625 and 626 tuners.

COMPONENT LOCATION - UHF TUNER KTT-626

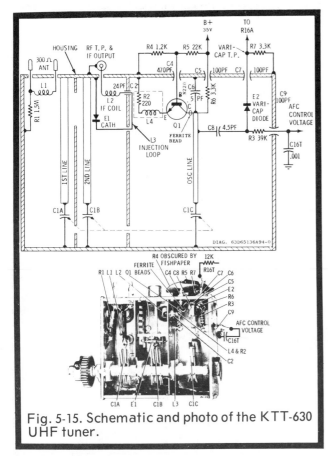

Fig. 5-15. Schematic and photo of the KTT-630 UHF tuner.

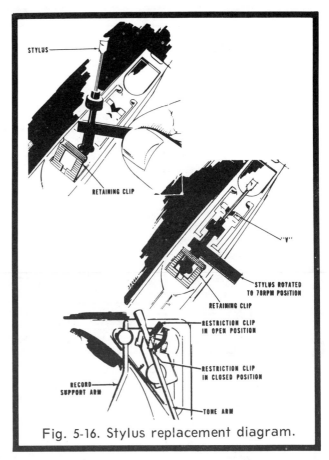

Fig. 5-16. Stylus replacement diagram.

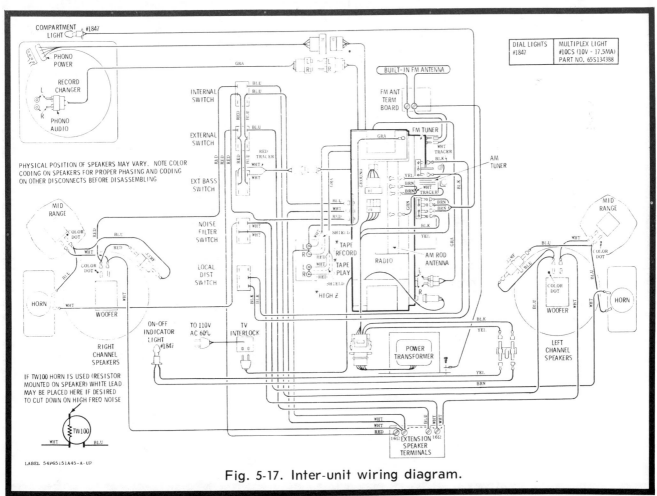

Fig. 5-17. Inter-unit wiring diagram.

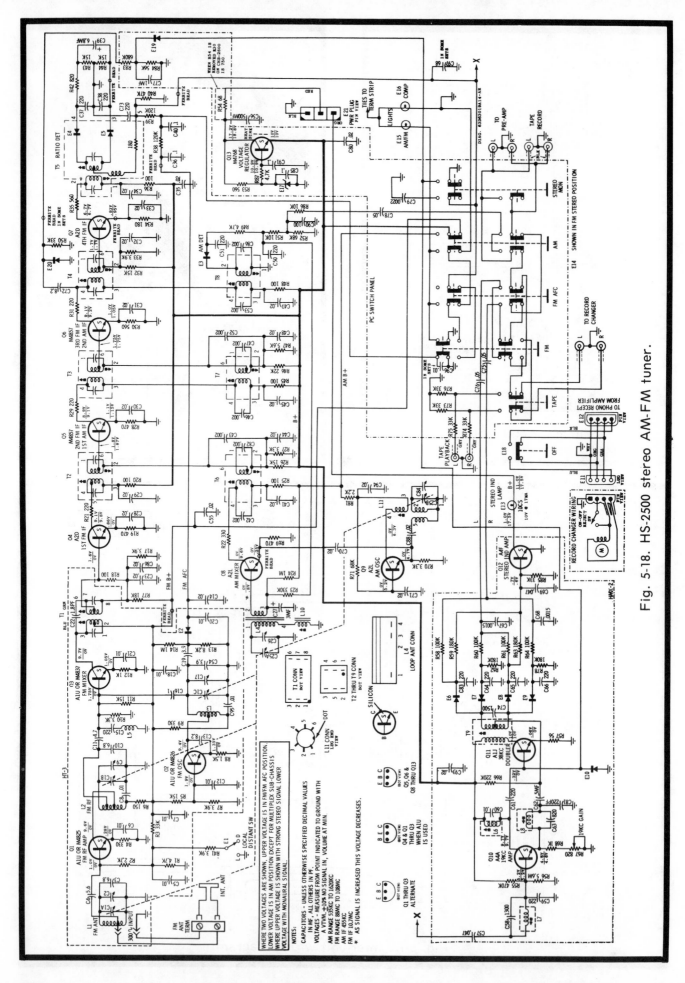

Fig. 5-18. HS-2500 stereo AM-FM tuner.

146

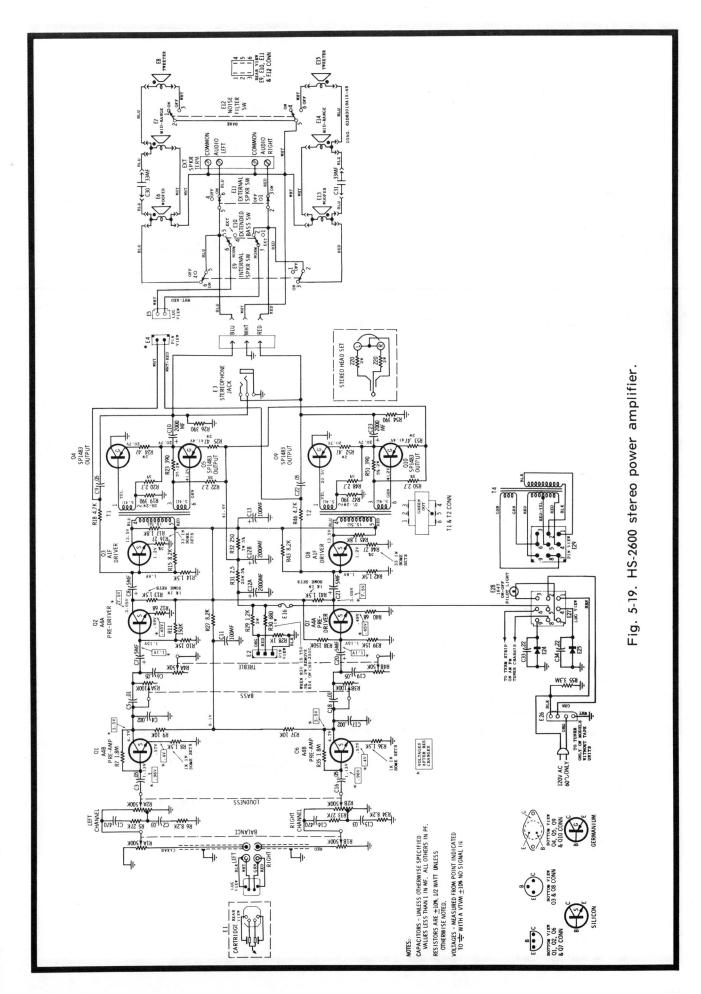

Fig. 5-19. HS-2600 stereo power amplifier.

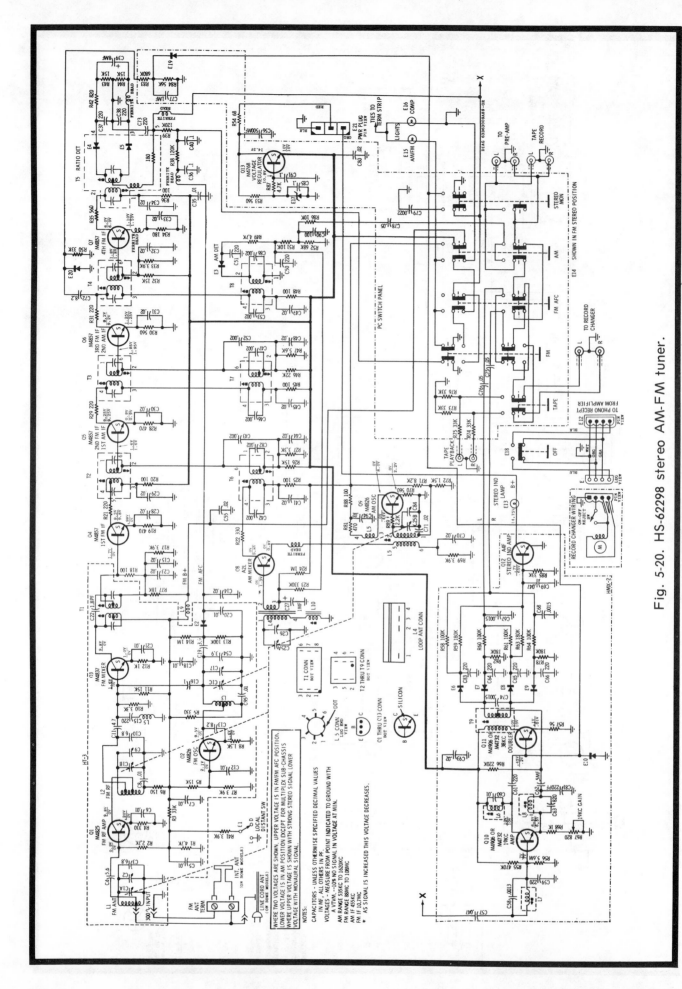

Fig. 5-20. HS-62298 stereo AM-FM tuner.

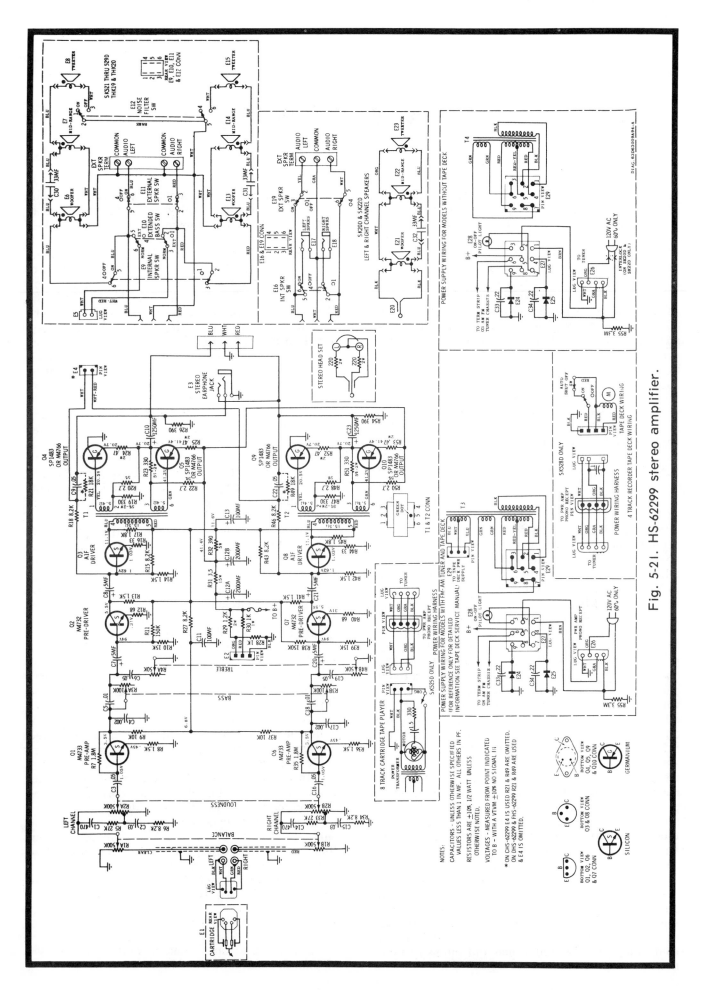

Fig. 5-21. HS-62299 stereo amplifier.

149

Remote Control Systems

In the Quasar and Quasar II series, Motorola actually uses one basic remote control in two slightly differing versions—the TRR-7 and the TRR-7Q series. In fact, the schematics and block diagrams are virtually identical; the remote hand oscillator is the same, but the Y panel has some minor changes, including a switch to make the early and late versions compatible. Because of Instamatic, the new TRR-11 remote is quite different, and actually has but three functions: off-on, channel selection, and audio control. The other functions provided by the previous systems are taken care of automatically by the AFT, demodulator phase-angle shift, and automatic intensity, which are all contained in the Instamatic operation. Volume stepping is done by a digital flip-flop on the remote chassis board of the receiver, which channels are selected conventionally by a motor.

TRR-7

The remote system (Fig. 6-1) used in the AC23TS-915 chassis consists of five functions: hue, intensity, volume, on-off and channel change. These remote functions are controlled by a battery-operated hand transmitter (Fig. 6-2) which employs a Hartley oscillator emitting selected CW signals at 35.5 kHz, 37.0 kHz, 38.5 kHz, 40.0 kHz, 41.5 kHz, 43.0 kHz and 44.5 kHz.

The input of the remote receiver contains a 3-transistor preamplifier stage (panel U) for selective gain of these signals. The output from the preamp is fed into an all-function driver (Q1X). The channel-change frequency (43-kHz) is fed through channel-change relay T2X and into amplifier Q2X which activates channel-change relay E1X. All other signals are diverted through function output amplifier Q1Y. These signals are applied to six high Q frequency discriminator coils (T1Y through T6Y) wired in series and respond only to the specific frequency selected by the hand transmitter for a desired function. The signal is then detected to produce a negative or positive control voltage for memory modules (E19Y, E20Y and E21Y). These modules contain a neon bulb, a high-grade capacitor and field-effect transistor. The control voltage causes the neon bulb to fire, charging the capacitor (C7Y, C20Y or C33Y). The charge voltage on these capacitors is used to control the amount of conduction of the FET (E14Y, E16Y and E18Y) memory modules. The source voltage developed across the resistors (R9Y, R31Y or R54Y) is the controlling bias for audio, hue and intensity functions.

For example, to turn the set on remotely and increase the volume (follow schematic diagram in Fig. 6-2), the hand transmitter emits a 38.5-kHz signal This signal is amplified in preamp panel U and fed to all-function driver Q1X, is amplified and fed to function output amplifier Q1Y. The output of Q1Y is fed to the six discriminator coils but only coil T2Y will respond. This signal is then detected by diode E2Y and fed to neon bulb E13Y in the memory module. When the voltage is sufficient (approximately 80v), the neon bulb fires, charging capacitor C7Y which forward biases the FET transistor. As the conduction of the FET increases, the voltage at the source resistor (R9Y) also increases. When this voltage reaches approximately 2v, the on-off relay amplifier Q3X is turned on and drives relay driver Q4X which, in turn, energizes on-off relay E2X and turns the set on.

A further increase in control voltage at source resistor R9Y applies a forward bias to second audio control amplifier Q3Y, increasing the audio output. This voltage increase is proportional to the audio output level.

To reduce volume and-or turn the set off, the described function is reversed. The hand unit output is 44.5-kHz. Only discriminator coil T1Y will respond; voltage developed at memory module E19Y will be negative, thereby reducing volume and when source voltage at R9Y drops below 2v, the set will turn off.

The hue and intensity functions operate in the same manner. The only difference is in the control amplifiers.

Audio: Q2Y and Q3Y. The audio level is controlled by the audio amplification in Q3Y and replaces the conventional volume control.

Hue: Q4Y and Q5Y. The total emitter-to-collector resistance of Q4Y and Q5Y and associated circuitry is varied by their conduction and replaces the conventional hue control.

Intensity: Q6Y. The emitter-to-collector resistance is varied by its conduction and replaces the conventional intensity controls.

The manual operation (control panel) is the same as remote operation. The difference being that the control voltage is fed to the respective memory modules through the control switches on the control panel which derives the control voltage from the power supply panel X.

TRR-7 TROUBLESHOOTING

If a remote function is inoperative, a quick method to isolate the problem is jumper respective

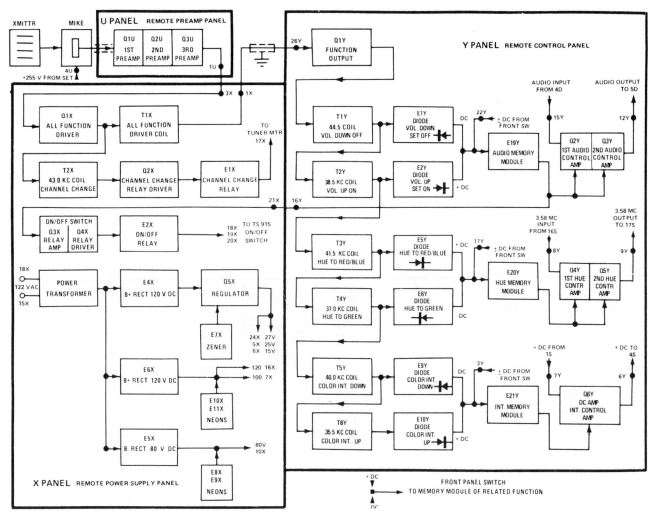

Fig. 6-1. Functional diagram, TRR-7 remote control.

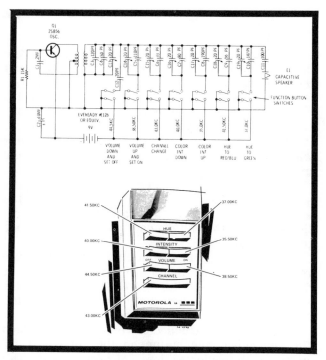

Fig. 6-2. Drawing and schematic of the TRR-7 hand transmitter.

terminals to by-pass suspected functions. If by placing a jumper between these terminals as indicated does not restore the function, the trouble is in the receiver, not the remote system. A slow operating function can be caused by leakage in diodes E3Y, E4Y, E7Y, E8Y, E11Y and E12Y. See Figs. 6-3 through 6-8.

1. Jumper 17X to 18X to turn the set on. Replace Q3X, Q4X, and on-off relay E2X if the jumper energizes the set but the panel won't. Alternate: Replace the X panel.

2. Check the four neon bulbs. Replace the X panel if any or all bulbs are not lit. Check the rectifier as an alternative (E4X, E5X and E6X).

3. Check the voltage at 3X. If not +1.5v DC, replace the X panel. Check and replace Q1X as an alternative.

4. Check the voltage at 3X. Push any function button. If the voltage does not drop below +1.5v DC, replace preamplifier panel U. Substitute a new microphone and try again. Pull the preamplifier panel and check the transistors.

5. Check the AC voltage at 1X. Push any function button. If the voltage does not rise, replace the X panel. Check or replace Q1X as an alternative. Range: No signal, 2v p-p; signal, 4v p-p.

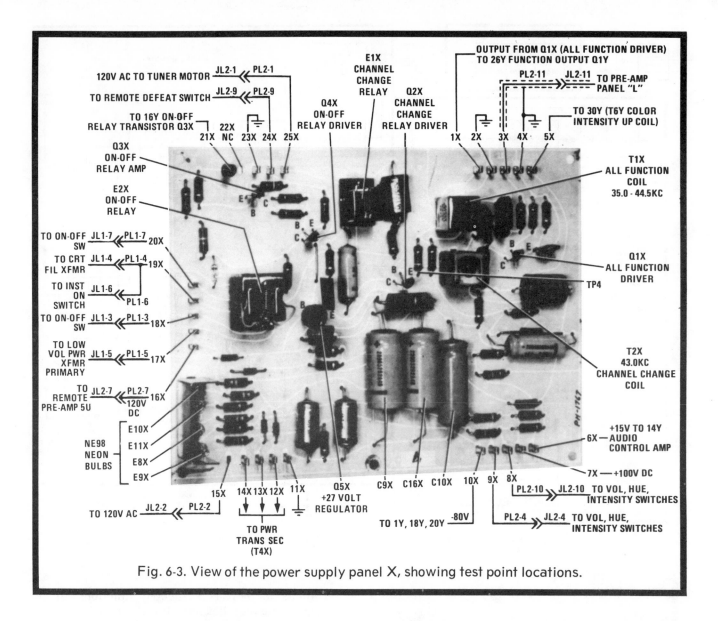

Fig. 6-3. View of the power supply panel X, showing test point locations.

6. Measure the collector voltage at Q1Y. Press any function button except channel change. If the voltage does not drop from about +30v DC to +20v DC or lower, change the Y panel. Check or replace Q1Y.

7. Bridge the in and out panel connectors of any suspected function to isolate a suspected function symptom to the receiver or remote system. Audio: 15Y to 12Y; Hu: 8Y to 9Y with a 1K potentiometer or a short; Intensity: 7Y to 6Y.

8. Check the cathode voltage of any suspected function diode (E1Y, E2Y, E5Y, E6Y, E9Y or E10Y). Press the related function button on the remote transmitter. If voltage does not go to +165v DC (plus or minus 25v DC), replace the diode. Check the discriminator coil.

9. Check the anode voltage of any suspected function diode. If the voltage does not go to -165v DC (plus or minus 25v DC), replace the diode. Check the discriminator coil.

10. Check the voltage at the junction of the diode load resistors (TP1, TP2 or TP3) of any suspected function. If the voltage does not swing to +60v DC with a positive output diode on, replace the diode.

Check the driving coil for the diode. If the voltage does not swing to -60v DC with a negative output diode on, replace the diode. Check the driving coil for that diode.

11. Check the source resistor voltage (R9Y, R31Y or R54Y) at any suspected memory module. If it does not vary smoothly between 0 and 15 volts when the function switch is depressed, replace the module.

12. Measure the collector voltage at any suspected function amplifier. (Note: Measure the emitter voltage of Q5Y.) Operate the related function switch. If the voltage does not change, replace the transistor.

13. Close the contacts on E1X (channel-change relay). If the tuner motor doesn't run, check the motor continuity and the AC to it. Check the motor and skip the switches if the motor stops immediately when the contacts are released.

14. Measure the collector voltage at Q2X. Push the channel-change button on the remote transmitter. If the collector does not drop to +10v DC, replace Q2X. Check T2X for continuity. Check the AC signal at the base of Q2X when the channel-

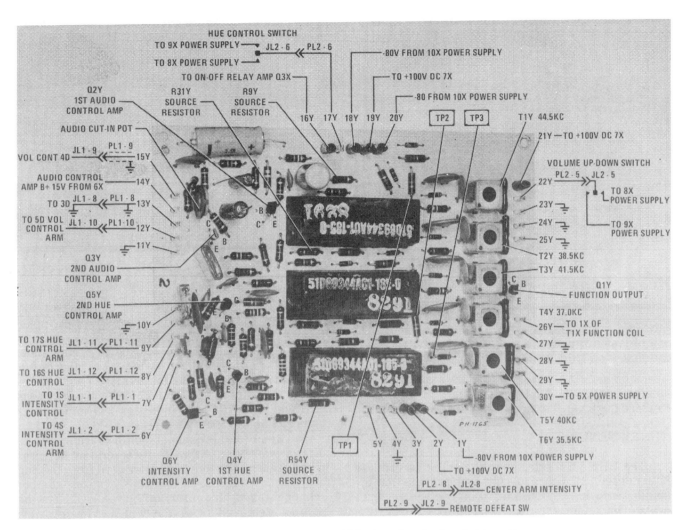

Fig. 6-4. View of remote panel Y, showing test point locations.

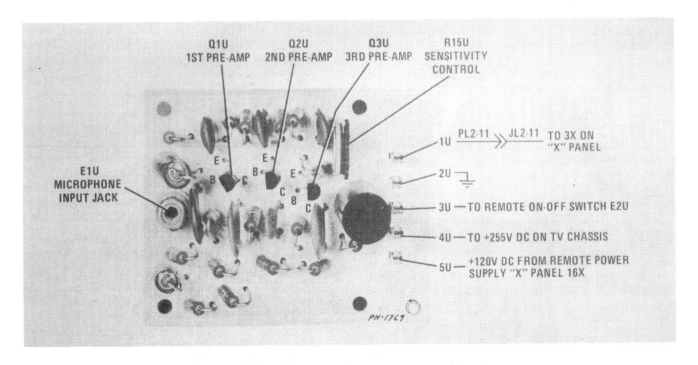

Fig. 6-5. View of the remote preamp panel U.

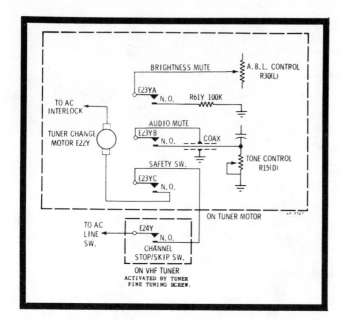

Fig. 6-6. Simplified wiring diagram of the brightness mute, audio mute, safety switch and channel stop-skip switch.

change button on the transmitter is pressed. Replace the Y panel if the signal is not present.

TUNER SETUP

The UHF-VHF selector cam can be turned in either direction with a plastic tool provided or a small screwdriver. Turning this control so the orange indicator arrow is pointing up (VHF position) allows the switching circuits to provide voltage to the VHF tuner. When the orange indicator arrow is turned down, UHF channels are selected for that position. The UHF tuner panel A (Fig. 6-9) receives voltage by means of the switching operated by the selector cam position.

VHF Tuner Preset Procedure

1. Turn the manual channel selector to the lowest VHF channel available.
2. Turn the VHF-UHF selector cam to VHF.
3. Push the AFT button to off (push-push switch).
4. Adjust the fine-tuning control. Turn the control to interference (sound bars in picture), then back off on the control to the clear picture. Repeat Steps 2, 3 and 4 for each VHF channel in the area.
5. Push the AFT button to on.

If no UHF channels are available, manually turn the channel selector knob to unused VHF channels and turn the fine-tuning control six or more turns counterclockwise. The tuner will automatically skip these positions when in remote operation or when the channel select button is depressed.

UHF Tuner Preset Procedure

1. Turn the manual channel selector to an unused channel.

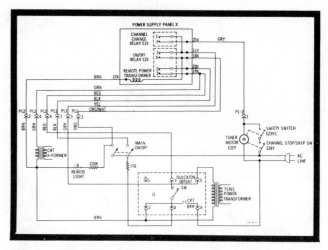

Fig. 6-7. Remote chassis AC wiring diagram.

2. Adjust the VHF-UHF selector cam to UHF.
3. Push the AFT switch to the off position (push-push switch).
4. Push in and turn the VHF-UHF tuning knob. A plastic tool or small screwdriver will aid in turning the control. Tune the lowest UHF channel; the tuning indicator arrow should be near the lower end of travel. The UHF dial indicator will move up as higher channels are tuned.
5. Remove the channel number from the indicator window; use fingers, tweezers, or longnose pliers. Take the channel indicator card from the bag at rear of set, punch out the desired number (channel tuned) and insert in the window. Save the number removed from the window for possible future use.
6. Repeat Steps 1, 2, 3, 4 and 5 for any other UHF station available.
7. Push the AFT knob to on.
8. If any channel is not used, turn the fine-tuning control six or more turns counterclockwise. The tuner will not stop at these positions when used in remote operation or when the channel select button is depressed.

TRR-7Q

Like the TRR-7, the TRR-7Q system performs five functions: hue, intensity, volume, on-off, and channel change. The only operational difference lies in the hue function, in which the source voltage from the hue memory module is applied as variable bias to hue output Q7. The voltage across the Q7 emitter resistor biases phase shifter Q13 on the color video panel and replaces the conventional hue control.

TRR-7Q TROUBLESHOOTING

The TRR-7Q service procedure is similar to that suggested for the TRR-7. However, several steps are different and they follow. For the steps omitted below, refer to the TRR-7 procedure.

7. Put a VTVM on the 500v scale. Connect the ground lead to TP1, 2 or 3. Connect the probe to the cathode of the up function diode E2Y, E5Y or E10Y.

INOPERATIVE FUNCTION	JUMPER CONTACTS	REPLACE OR CHECK
Set will not turn on	17X and 18X	On-Off relay E2X and associated wiring
Loss of Audio	12Y and 15Y for maximum audio	Q2Y, Q3Y and associated components
Loss of Intensity	6Y and 7Y for maximum intensity	Q6Y and associated components
Loss of Hue	8Y and 9Y for maximum magneta	Q4Y, Q5Y and associated components

Table 6-1. TRR-7 remote control troubleshooting chart.

Press the related function button on the remote transmitter. If the voltage does not go to 150v or more, replace the diode. Check the discriminator coil.

8. With the ground lead of the VTVM on TP1, 2 or 3, put the probe on ground and reverse the polarity of the meter. Press the related down function button on the remote transmitter. If the voltage does not go to -150v or more, replace the down function diode D1Y, E6Y or E9Y. Check the discriminator coil.

9. Bridge the in and out panel connectors of any suspected function to isolate the suspected function symptom to the receiver or remote system. Audio: 15Y to 12Y; Hue: emitter of Q7Y (white lead) to ground; Intensity: 7Y to 6Y with a 10K ½w resistor.

10. Check the voltage at the memory module input. In manual operation the input voltage should be plus or minus 40v. If not, check resistors R8Y, R29Y or R51Y. Also check the manual function switch for plus or minus voltage. In remote operation, the input voltage should be plus or minus 20v. If not, check resistors R73Y, R73Y or R75Y.

12. Measure the collector voltage at any suspected function amplifier. (Note: Measure the emitter voltage of Q7Y.) Operate the related function switch. If the voltage does not change, replace the transistor.

Tuner setup is the same as listed for the TRR-7.

ALIGNMENT PROCEDURE, TRR-7 & TRR-7Q

Alignment is a matter of using a scope or DC voltmeter as an indicator connected to the output of a function diode (Test Points 1, 2 and 3 on the Y panel), and adjusting the appropriate transmitter trimmer for maximum signal at the test points.

Coil T1X is a broadband coil and is tuned for maximum at 40 kHz center frequency. Since the intensity-down coil, T5Y, is tuned to this frequency also, the intensity memory module output voltage can be used as an indicator.

The channel change coil, T2X, can be tuned to 43 kHz while watching the channel-change relay. The relay will energize when the coil is tuned to 43 kHz. Tune through the coil's range to determine when the

relay pulls in and drops out. Leave the coil slug centered between these two points.

To align the transmitter, use a receiver that is known to be properly aligned. Adjust the trimmer capacitors as indicated in the alignment chart (Table 6-2). To align the receiver, use a transmitter that is known to be properly aligned. Adjust the coils on the X and Y panels as indicated on the chart.

TRR-11

This remote control system (1) turns the TV receiver on and off, (2) varies the volume of the audio in three steps, with a lamp to indicate the volume level, and (3) changes channels. In addition to the remote control, these same functions are provided by pushbuttons on the receiver front panel.

The receiver is turned on by depressing either the on-off volume button on the remote control transmitter or the on-off volume knob on the front panel. The volume level is increased and the indicator lamp brightness varied in three steps: 1) low volume level, indicator lamp is dim; 2) medium volume level, indicator lamp is brighter and; 3) high volume level, indicator lamp is brightest. Volume can also be adjusted by rotating the volume control knob on the front panel. Depressing this knob changes the volume level and indicator lamp brightness the same as with remote operation. Channel Change is accomplished by remote control or by a front panel power-tune button. Sound and picture are electronically muted during tuner rotation.

A separate transformer supplies power to the remote control panel YA and filament voltage to the picture tube at a reduced level with the receiver off. The transformer is physically located at the rear of the chassis, near vertical deflection panel VA. Most separate plug in panel, YA.

Remote Transmitter

The remote control transmitter has two pushbuttons. Depressing a button causes the spring-loaded hammer to strike the end of a mechanically resonant rod. Vibrations travel through the length

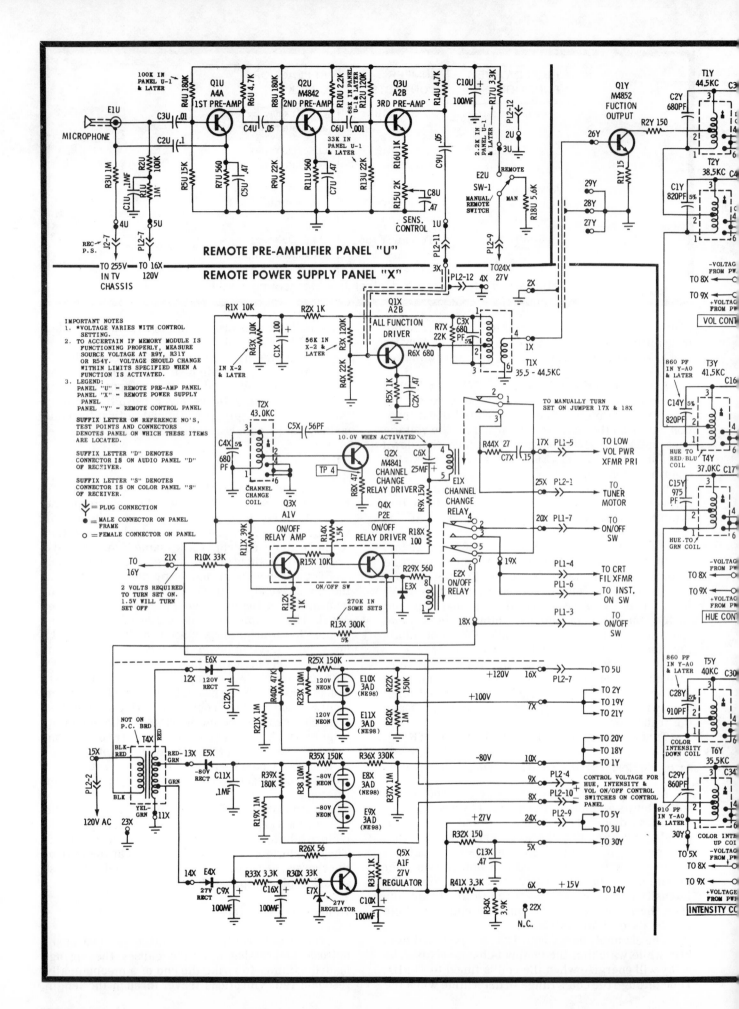

REMOTE PRE-AMPLIFIER PANEL "U"

REMOTE POWER SUPPLY PANEL "X"

IMPORTANT NOTES
1. *VOLTAGE VARIES WITH CONTROL SETTING.
2. TO ACCERTAIN IF MEMORY MODULE IS FUNCTIONING PROPERLY, MEASURE SOURCE VOLTAGE AT R9Y, R31Y OR R54Y. VOLTAGE SHOULD CHANGE WITHIN LIMITS SPECIFIED WHEN A FUNCTION IS ACTIVATED.
3. LEGEND:
PANEL "U" – REMOTE PRE-AMP PANEL
PANEL "X" – REMOTE POWER SUPPLY PANEL
PANEL "Y" – REMOTE CONTROL PANEL

SUFFIX LETTER ON REFERENCE NO'S, TEST POINTS AND CONNECTORS DENOTES PANEL ON WHICH THESE ITEMS ARE LOCATED.

SUFFIX LETTER "D" DENOTES CONNECTOR I8 ON AUDIO PANEL "D" OF RECEIVER.

SUFFIX LETTER "S" DENOTES CONNECTOR I8 ON COLOR PANEL "S" OF RECEIVER.

= PLUG CONNECTION
= MALE CONNECTOR ON PANEL FRAME
= FEMALE CONNECTOR ON PANEL

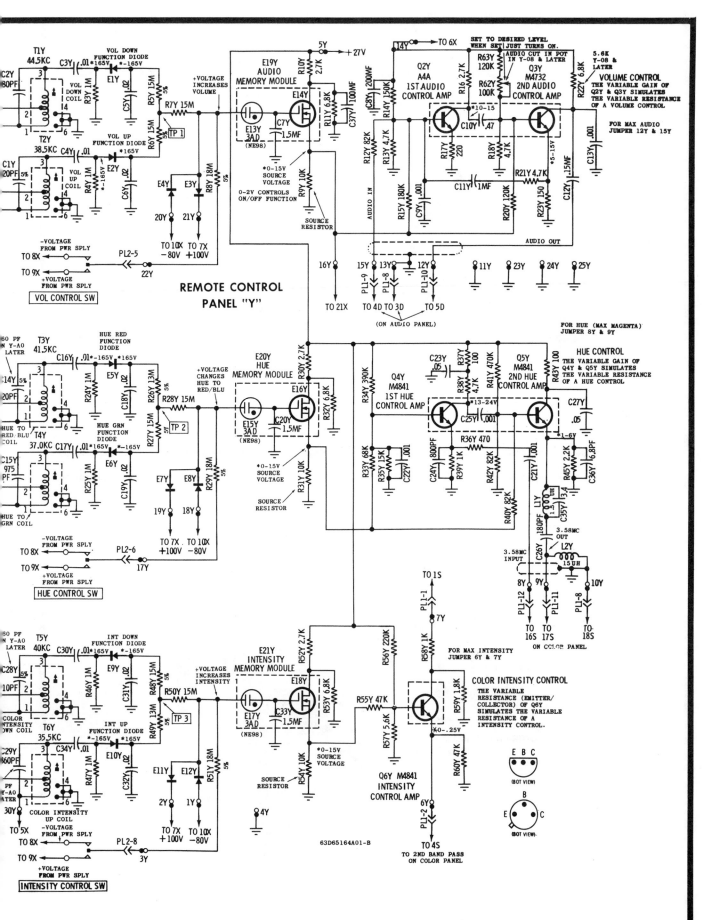

Fig. 6-8. Schematic of the TRR-7 remote system.

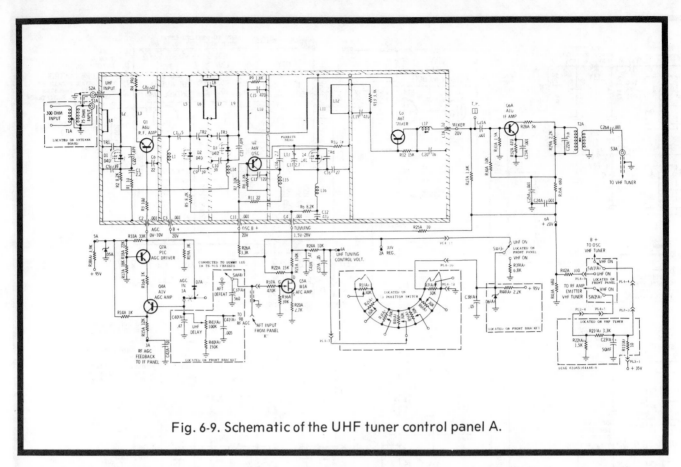

Fig. 6-9. Schematic of the UHF tuner control panel A.

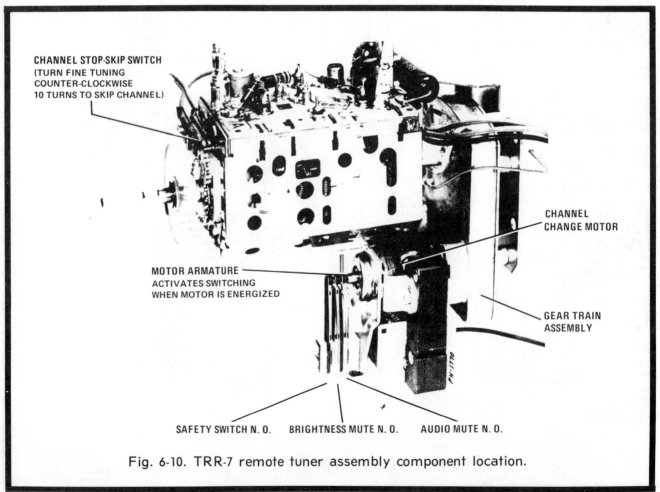

CHANNEL STOP-SKIP SWITCH
(TURN FINE TUNING
COUNTER-CLOCKWISE
10 TURNS TO SKIP CHANNEL)

**CHANNEL
CHANGE MOTOR**

MOTOR ARMATURE
ACTIVATES SWITCHING
WHEN MOTOR IS ENERGIZED

**GEAR TRAIN
ASSEMBLY**

SAFETY SWITCH N. O.   BRIGHTNESS MUTE N. O.   AUDIO MUTE N. O.

Fig. 6-10. TRR-7 remote tuner assembly component location.

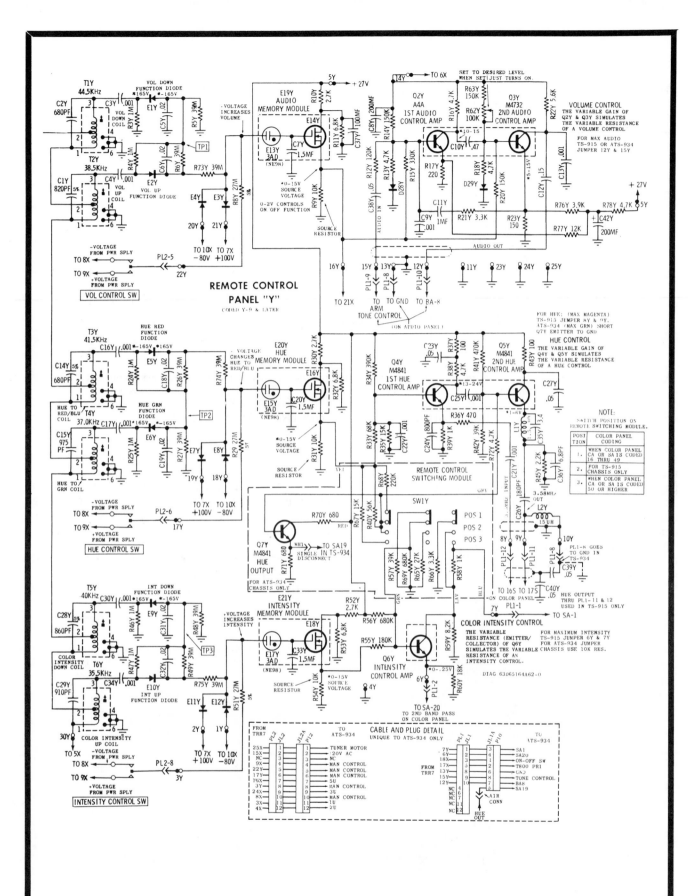

Fig. 6-11. Schematic of the TRR-7Q Y panel. Most circuits are identical with the TRR-7, except hue control.

TO ALIGN "Y" PANEL COILS	TO ALIGN "X" PANEL COILS	TO ALIGN TRANSMITTER TRIMMERS	CONNECT VTVM ON "Y" PANEL GND LEAD TO:	CONNECT VTVM ON "Y" PANEL PROBE TO:	CONNECT VTVM TO T.P. ON "X" PANEL	ACTIVATE TRANSMITTER FUNCTION	ADJUST FOR
Coil T3Y	—	C18 (41.5Kc)	TP2	E5Y cathode	—	Hue to Red	Maximum positive
Coil T4Y	—	C19 (37Kc)	TP2	Gnd	—	Hue to Green	
Coil T6Y	—	C17 (35Kc)	TP3	E10Y cathode	—	Intensity Up	
Coil T5Y	—	C7 (40Kc)	TP3	Gnd	—	Intensity Down	
Coil T2Y	—	C14 (38.5Kc)	TP1	E2Y cathode	—	Volume Up/On	
Coil T1Y	—	C13 (44.5Kc)	TP1	Gnd	—	Volume Down/Off	
—		C15 (43Kc)	—	—	TP4: R8X to ground	Channel Change	
—	Coil T1X	—	TP3	Gnd	—	Intensity Down	
—	Coil T2X	—	—	—	TP4: R8X to ground	Channel Change	

NOTE: Reduce sensitivity control on pre-amp panel "U" or position transmitter further away from receiver to prevent overloading and broad tuning.

Table 6-2. Alignment chart for TRR-7 and TRR-7Q remote systems.

of the rod, setting up oscillations within it at the mechanical resonant frequency, determined by its physical length. The rod is one-half wavelength long and the intensity of vibrations at the hammer end is approximately equal at the opposite end. At the center of the length of rod a vibration null exists. This makes a convenient point to hold or suspend the rod without damping the oscillation. The rod acts as a transducer converting mechanical energy—vibrations within the rod—to acoustical energy.

One transducer emits energy at 41.5 kHz and activates the remote receiver channel-change circuits. The larger transducer emits energy at 38.5 kHz and activates the remote receiver on-off and volume step circuits. Each transducer, after being struck, is damped by the hammer when the push-button is released and only a short burst of acoustical energy is emitted for each hammer blow.

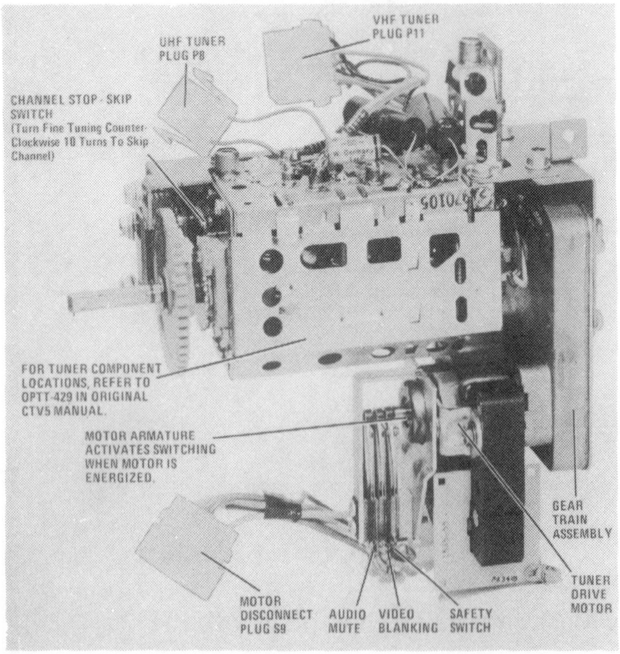

Remote Receiver

When the channel-change function button on the remote control transmitter is activated, the remote

Fig. 6-12. TRR-7Q remote tuner assembly, TRR-7Q.

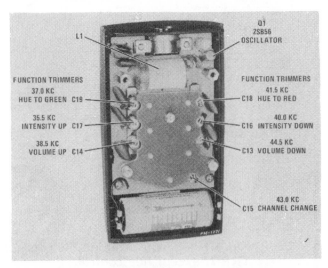

Fig. 6-13. Disassembled view of the hand transmitter.

Fig. 6-14. TRR-11 remote control transmitter.

receiver microphone receives the 41.5-kHz signal and passes it to the preamplifier. The amplified signal is applied to the channel-change pulse detector. When this amplifier conducts, the relay operates and closes two contacts, applying 120v AC to the channel-change tuner motor and causes the motor to rotate and change channels.

Three NPN transistors, Q1, Q2 and Q3 (preamplifier, Fig. 6-16), amplify the output from the crystal microphone. Direct-coupled common-emitter circuits are utilized. The sensitivity of the amplifier is determined by variable control R10

(10K) between the collector of Q3 and ground. The output of the preamp is capacitance coupled (1) to the channel change circuits and (2) to the on-off volume and lamp level circuits.

The 41.5-kHz signal from the preamplifier is capacitance coupled (Fig. 6-16) by C12 (56 pf) to a tuned circuit (L2, C13). This high Q frequency selective circuit is tuned by coil L2 and selects the proper signal. The relay winding (RE1) is the collector load for the channel-change pulse detector (Q4). With the proper signal applied, the transistor conducts through the relay winding and operates the relay, closing its contacts. AC line voltage is applied to the channel-change motor winding. The tuner motor armature is drawn into the motor housing, activating the mute switch mounted at the rear of the armature. The mute circuit blanks the picture tube and mutes the audio during tuner rotation.

The Q4 current flows through emitter resistor R41, common to both detectors (channel-change and off-on vol.) and develops a negative-going (less positive) voltage. This voltage is applied to the emitter of Q5 (off-on, volume detector) and biases it off to ensure operation of only one function at a time. The mute switch applies a positive voltage by voltage divider R44 and R46 to Terminal 17TA, IC color demodulator, and the picture tube is blanked during channel change. It also forward biases diode D3, muting the sound. The diode is connected to the high side of the volume control and is in series with resistor R48 (100-ohm) to ground. The volume

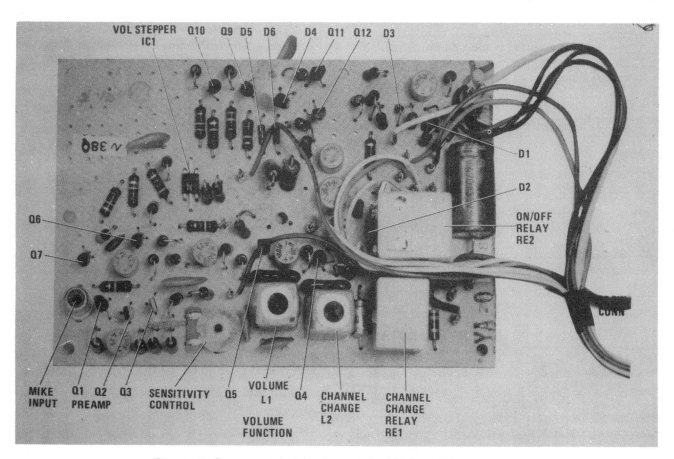

Fig. 6-15. Component side view of the TRR-11 YA panel.

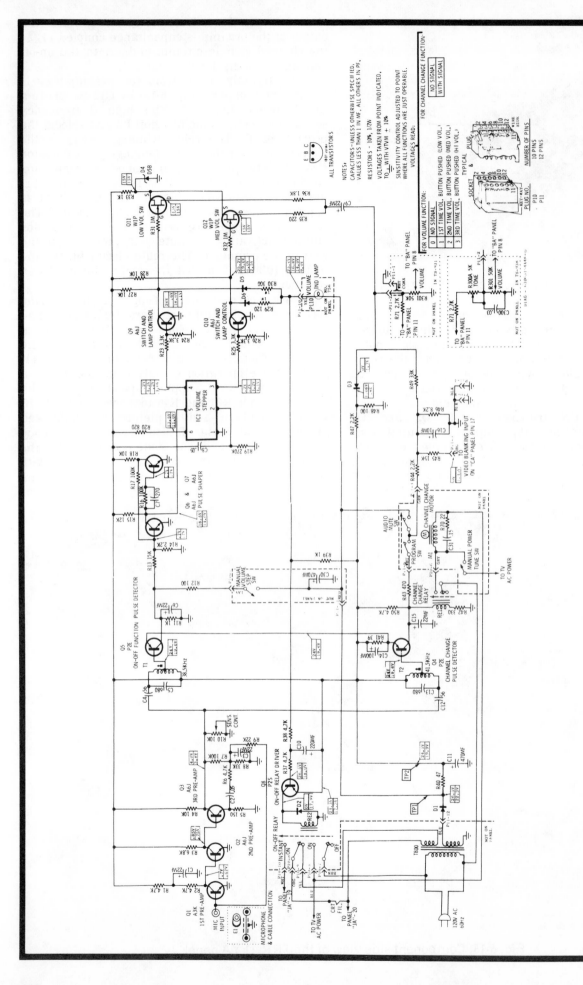

Fig. 6-16. Schematic of the TRR-11 receiver.

control is shunted to ground, muting the audio by the low forward resistance of D3 and R48. When the tuner stops rotating the mute switch is opened. Resistors R47 and R48 form a voltage divider from B+ and diode D3 is reverse biased, allowing the volume to return to normal level. To skip undesired channels, the fine-tuning control is turned counterclockwise (to the left) about ten turns.

When the on-off volume button on the remote transmitter is depressed, acoustical energy is radiated and received by the microphone. Following preamplification, the signal is detected by the volume pulse detector and is directed to the pulse shaper. The sensitivity control varies the amount of signal available and the range of remote operation. The pulse shaper (Fig. 6-16) shapes the pulse (steep skirts) which activates the volume stepper, consisting of two bistable multivibrators and control circuits in one package (integrated circuit).

The TV volume control receives the audio signal at the high side of the control. Audio is removed by the control arm and fed to the audio amplifier circuit. A large capacitance is connected to the high side of the control, and in series with the capacitance are two parallel resistors (220 and 1.8K ohms). The resistors are returned to ground by two closed switches (transistors), placing the resistors in parallel with the volume control. With the transmitter volume function pushbutton depressed the first time, control voltage from the stepper activates the low volume level channel. This opens the switch and removes the 1.8K resistor from the circuit, leaving the volume control shunted by the 220-ohm resistor, resulting in low volume. The indicator lamp is lighted dimly through the 300-ohm resistor and the indicator lamp switch.

When the transmitter pushbutton is depressed the second time, the low volume level channel returns to its original state, and the medium volume level channel is activated. The 220-ohm resistor is removed from the circuit by the open switch and the volume control is shunted by the 1.8K resistor, providing increased volume. The indicator lamp lights brighter through the 120-ohm resistor and it's associated lamp switch.

When the transmitter volume pushbutton is depressed the third time, both low and medium volume level channels are activated. Both of the resistors are removed from the circuit by the open switches, resulting in high volume. The indicator lamp is now brilliantly lighted through the 120-ohm resistor and its switch, and the 300-ohm resistor and its switch. The fourth transmitted pulse returns both low and medium volume channels to their original state. Subsequent transmitted pulses repeat the above mentioned actions.

When the on-off volume button on the remote transmitter is depressed, a 38.5-kHz signal is received by the microphone and amplified by the preamplifier. The output of the preamp is applied to the base of the on-off volume pulse detector (Q5) by coupling capacitor C4 and high Q resonant circuit (L1, C5). The signal is detected and current flows through the common-emitter resistor, biasing off the channel-change detector to ensure operation of the selected function.

The positive pulse of voltage developed at the Q5 collector is applied to a pulse-shaper circuit.

The pulse shaper is a flip-flop circuit consisting of transistors Q6 and Q7. Resistors R15 and R16 apply forward bias to the base of Q7 and it conducts heavily. The low collector voltage of Q7 is coupled to the base of Q6 by a voltage divider (R17 and R14) and Q6 does not conduct. This is the normal stable condition—Q6 turned off and Q7 conducting heavily.

When the positive pulse is applied to the base of Q6 (from the pulse detector), conducting occurs, resulting in a lower collector voltage. This decrease of voltage is applied to Q7 by R16 and C7, rapidly biasing Q7 to cutoff. Q6 is driven to heavy conduction, aided by the feedback applied by R17. Q6 cannot remain conducting without a trigger pulse and Q6 and Q7 revert to their original stable state. The output of the shaper circuit is a negative-going square wave with very short fall time (steep skirt).

A manual on-off volume step switch is mounted on the front panel of the receiver, which allows manual operation of the on-off volume in the same sequence as the remote transmitter. When SW1 (part of R301) is momentarily closed, a pulse is formed from B+ to a voltage divider consisting of R12 (100-ohm) and R11 (1K). This pulse is further divided by R13 (15K) and R14 (2.2K), and it triggers the shaper.

The negative voltage pulse from the pulse shaper is applied to a volume-stepper circuit consisting of two bistable multivibrators and control circuits connected in cascade. Control voltages generated by the volume stepper are applied to the lamp control and volume switch circuits.

With either or both of the lamp control switches activated, the on-off relay driver is turned on. The relay contacts close and AC is applied to the voltage-doubler circuit to supply power to the switch mode power supply. CRT filament power is switched from the remote control transformer to the switch mode power supply.

PNP transistor Q8 is a relay driver for the on-off function. A relay winding (RE2) is the collector load. The emitter is returned to B+. The base is coupled through an RC time constant (R37, R38 and C10) to the indicator lamp and its shunt resistor R39 (1K). With a pulse from the transmitter or front panel on-off volume switch, Q9, Q10 or both conduct and reduce the positive voltage at the base or Q8, forward biasing it. Current flowing through relay winding RE2 activates the relay coil and closes its contacts. By closing the relay contacts, the circuit is completed from the 120v AC source to the receiver voltage doubler input and the set is turned on. The RC time constant in the base circuit of Q8 allows about 6 seconds before turning the receiver off. This allows passing from a high volume level through off and back to on with a low level of volume without interruption.

Parts Lists

The following parts lists include coils, transformers, special transistor information, critical hardware and other special components. Omitted are common capacitors, resistors, etc. When replacing capacitors on the video driver and other color panels, you'll notice that a number of NPO (no parts per million) capacitors are used. Be absolutely sure you substitute identical units; otherwise, you can expect significant signal drift.

TS915-919 QUASARS

REF.	PART NO.	DESCRIPTION	REF.	PART NO.	DESCRIPTION
		VIDEO IF PANEL "B"			**TRANSISTORS**
			Q-3D	48S134733	1ST AUDIO AMP (4733): use A3K
		ELECTRICAL PARTS		48S137015	1ST AUDIO AMP A4B: use A3K
				48S134997	1ST AUDIO AMP A3K: in panel D-3C
		MISCELLANEOUS ELECTRICAL PARTS	Q-4D	48S134732	AUDIO DRIVER (4732): use A3K
				48S137014	AUDIO DRIVER A4A: use A3K
E-1B	48C65837A02	DIODE, crystal		48S134997	AUDIO DRIVER A3K: in panel D-3C
E-2B	48B66629A05	DIODE, low power			
E-3B	48S137000	DIODE, zener: D1S in panel B-8A			**TRANSFORMERS**
			T-1D	24D68772A03	AUTO TRANSFORMER
		COILS & CHOKES	T-2D	24D10282A02	RATIO DETECTOR
L-1B	24D68501A01	CONVERGENCE SEC			
L-2B	24D68501A02	47.25Mc TRAP			**MECHANICAL PARTS**
L-3B	24D68501A03	39.75Mc TRAP		29S10134A02	CONNECTOR, recept (PLAcir chassis mtg - 10 used)
L-4B	24D68501A03	35.25Mc TRAP		9C67349A03	RECEPTACLE, phono
L-5B	24D67676A19	COMPENSATING: 15 uh			
L-6B	24D66772A12	RESONANT CHOKE: 7.5 uh			
L-7B	24V68614A76	4.5Mc TRAP: incls C-35B & R-51B			**VIDEO PRE-AMPLIFIER PANEL "E"**
L-8B	24D68501A07	41.25Mc PHASE ADJ			
L-9B	24D67676A20	COMPENSATING: 450 uh			ELECTRICAL PARTS
L-10B	24D66772A12	RESONANT CHOKE: 7.5 uh			
L-11B	24D66772A12	CHOKE: 7.5 uh in panel B-6			MISCELLANEOUS ELECTRICAL PARTS
			E-1E	48D67120A02	DIODE, low power
		TRANSISTORS			
					COILS
Q-1B	48S134981	1ST IF: A2Y; use A1G or A1G-1	L-2E	24D68801A03	COMPENSATING: 100 uh
	48S134904	1ST IF: A1G or A1G-1 (replace with same type or in pairs)	L-3E	24D68852A02	DELAY LINE
Q-2B	48S134981	2ND IF: A2Y; use A1G or A1G-1	L-4E	24D68801A22	COMPENSATING: 200 uh
	48S134904	2ND IF: A1G or A1G-1 (replace with same type or in pairs)	L-5E	24D67676A21	COMPENSATING: 100 uh
Q-3B	48S134937	3RD IF: A1Z	L-6E	24D66772A12	IF RESONANT
Q-4B	48S134910	AGC AMP: P1C	L-9E	24D66772A12	IF RESONANT
Q-5B	48S134841	1ST DETECTOR: 4841			
Q-6B	48S134841	VIDEO DETECTOR OUTPUT AMP: 4841			**TRANSISTORS**
			Q-1E	48S134841	1ST VIDEO (4841)
		CONTROLS	Q-2E	48S134910	2ND VIDEO (P1C)
R-1B	18D66401A22	47.25 TRAP ADJUST: 100Ω	Q-3E	48S134910	NOISE SEPARATOR (P1C)
R-28B	18D66401A25	RF DELAY: 3.5K	Q-4E	48S134842	NOISE INVERTER (4842)
	18D66401A29	RF DELAY: 25K in panel B-1A & B-3	Q-5E	48S134842	AGC GATE (4842)
		TRANSFORMERS	Q-6E	48S134933	SYNC SEPARATOR (A1V)
T-1B	24D68501A04	IF INTERSTAGE	Q-7E	48S134841	BRIGHTNESS LIMITER (4841)
T-2B	24D68501A04	2ND IF INTERSTAGE		48S134953	BRIGHTNESS LIMITER: A2K in panel E-6A & later
T-3BA	24D68501A05	3RD IF PRIMARY	Q-8E	48S134842	SYNC INVERTER (VERTICAL (4842)
T-3BB	24D68501A06	3RD IF SECONDARY	Q-9E	48S134952	VERTICAL MULTI-VIBRATOR (A2J)
T-4B	24D68501A08	41.25Mc TRAP	Q-10E	48S134943	VERTICAL SHAPER (P1J)
			Q-11E	48S134933	VERTICAL DRIVER (A1V)
		MECHANICAL PARTS			
	29S10134A20	CONNECTOR, recept (for PC mtg - 10 required)			CONTROLS
	29S10134A04	PIN, terminal (2 used)	R-13E	18D66401A24	NOISE GATE 10K
	9C67349A04	RECEPTACLE, 1 pin	R-24E	18D66401A23	AGC THRESHOLD 1K
		IC AUDIO PANEL "D"			MECHANICAL PARTS
		ELECTRICAL PARTS		42C68847A01	CLIP, plastic: delay line coil mtg
		MISCELLANEOUS ELECTRICAL PARTS		29S10134A11	CONNECTOR, recept: delay line coil (3used)
IC-1D	51S10276A01	INTEGRATED CIRCUIT (IC) T1A			
E-1D	48S134957	DIODE, silicon Zener D1G			

REF.	PART NO.	DESCRIPTION
	29S10134A20	CONNECTOR, recept: P.C. mtg (20 required)
	29S10134A04	PIN, terminal (used with 29S10134A11)
	26A66745A03	SINK, heat: Q-7E in panel E-6A & later
	43B68719A01	SPACER, transistor mtg: Q-7E in panel E-6A & later

HORIZONTAL OUTPUT PANEL "F"
ELECTRICAL PARTS

MISCELLANEOUS ELECTRICAL PARTS

REF.	PART NO.	DESCRIPTION
E-1F	48S134921	DIODE, silicon (D1D)
E-2F	48S134921	DIODE, silicon (D1D)
E-3F	48S134959	DIODE, silicon (D1J)
E-4F	48S134959	DIODE, silicon (D1J)
E-5F	76A544401	BEADS, ferrite
E-6F	76A544401	BEADS, ferrite

COILS

REF.	PART NO.	DESCRIPTION
L-1F	24D68778A01	HORIZONTAL OSCILLATOR
L-2F	24D68801A16	COMPENSATING: 6600 uh
L-3F	24D68573A01	HORIZONTAL SPOOK
L-4F	24D69267A01	BALANCING COIL in panel F-3A & F-5
	24D69267A02	BALANCING COIL

TRANSISTORS

REF.	PART NO.	DESCRIPTION
Q-1F	48S134917	DUAL DIODE, detector (D1C)
Q-2F	48S134842	HORIZONTAL OSCILLATOR (4842)
	48S137006	HORIZONTAL OSCILLATOR A3S in panel F-6
Q-3F	48S134815	HORIZONTAL PRE-DRIVER M4815 (in some panels)
	48S134910	HORIZONTAL PRE-DRIVER (P1C)
Q-4F	48S134919	HORIZONTAL DRIVER (A1M)
Q-5F	48S134910	ARC GATE (P1C)
Q-6F	48S134901	HORIZONTAL OUTPUT A1D: use A3H
	48S134995	HORIZONTAL OUTPUT A3H; in panel F-5
Q-7F	48S134901	HORIZONTAL OUTPUT A1D: use A3H
	48S134995	HORIZONTAL OUTPUT A3H: in panel F-5
Q-8F	48S134842	COLOR GATE DRIVER (4842)

CONTROLS

REF.	PART NO.	DESCRIPTION
R-23F	18D68447A03	HORIZONTAL CENTERING: 10Ω

TRANSFORMERS

REF.	PART NO.	DESCRIPTION
T-1F	25D68782A01	HORIZONTAL DRIVER

MECHANICAL PARTS

REF.	PART NO.	DESCRIPTION
	29S10134A20	CONNECTOR, receptacle (for PLAcir chassis mtg - 15 used)
	14C66650A02	INSULATOR, rod adjustment mtg
	47C69019A01	POST, plastic: heat sink support
	47C66082A02	ROD, adjustment (L-1F)
	26D68540A01	SHIELD, metal (for horizontal oscillator coil L-1F)
	26A68999A02	SINK, heat (for transistors Q-6F & Q-7F)
	9C63825A01	SOCKET, for transistors Q-6F & Q-7F

PINCUSHION PANEL "G"
ELECTRICAL PARTS

MISCELLANEOUS ELECTRICAL PARTS

REF.	PART NO.	DESCRIPTION
E-1G	48S191A08	DIODE, silicon in panel G-1A only)

COILS

REF.	PART NO.	DESCRIPTION
L-1G	24D68778A03	COIL, top tilt

TRANSISTORS

REF.	PART NO.	DESCRIPTION
Q-1G	48S134838	VERTICAL PIN AMP (4838)

REF.	PART NO.	DESCRIPTION
Q-2G	48S134842	REG DRIVER (4842)
Q-3G	48S134918	REG AMP (A1L)

TRANSFORMERS

REF.	PART NO.	DESCRIPTION
T-1G	25D68782A03	VERT PIN MOD

CONTROLS

REF.	PART NO.	DESCRIPTION
R-2G	18D68447A04	BOTTOM PIN AMP 50Ω
R-9G	18D66401A23	HORIZONTAL SIZE 1K
R-16G	18D66401A26	SIDE PIN 100K

MECHANICAL PARTS

REF.	PART NO.	DESCRIPTION
	29S10134A20	CONNECTOR, recept (10 required)
	26A66745A01	SINK, heat: for Q-1G transistor
	43B68719A01	SPACER: for Q-1G transistor

CONVERGENCE PANEL "H"
ELECTRICAL PARTS

MISCELLANEOUS ELECTRICAL PARTS

REF.	PART NO.	DESCRIPTION
E-1H	48S10062A01	RECTIFIER, silicon
E-2H	48S191A08	RECTIFIER, silicon
E-3H	48S191A08	RECTIFIER, silicon

COILS

REF.	PART NO.	DESCRIPTION
L-1H	24D67682A06	R-G HORIZONTAL LINES
L-2H	24D67682A03	BLUE TILT HORIZONTAL
L-3H	24D67682A03	BLUE CENTER HORIZONTAL PHASE
L-4H	24V68609A47	R-G RIGHT SIDE VERTICAL LINE

TRANSISTORS

REF.	PART NO.	DESCRIPTION
Q-2H	48S134842	CURRENT LIMITER (4842)

TRANSFORMERS

REF.	PART NO.	DESCRIPTION
T-1H	25D68538A01	VERTICAL OUTPUT

CONTROLS

REF.	PART NO.	DESCRIPTION
R-4H	18D66401A21	VERTICAL LINEARITY: 1K
R-8H	18D67671A01	VERTICAL SIZE: 30Ω
R-12H	18D67671A13	VERTICAL CENTERING: 10Ω
R-14H	18D67671A01	R-G VERT DIFF TILT: 30Ω
R-15H	18D67671A07	R-G VERT TILT: 200Ω
R-16H	18D67671A07	BLUE VERT TILT: 200Ω
R-17H	18D67671A07	R-G VERT DIFF AMP: 200Ω
R-18H	18D67671A07	R-G VERT AMP: 200Ω
R-19H	18D67671A01	BLUE VERTICAL AMP: 30Ω
R-22H	18D67671A14	R-G LEFT SIDE VERT LINES: 90Ω
R-23H	18D67671A11	HORIZ DIFF TILT: 150Ω
R-24H	18D67671A04	HORIZ BLUE AMP: 150Ω (USE 18D67671A11)

RESISTORS

REF.	PART NO.	DESCRIPTION
R-1H	6C66263A08	VARISTOR

MECHANICAL PARTS

REF.	PART NO.	DESCRIPTION
	29S10134A20	CONNECTOR, recept: PLAcir chassis mtg (15 required)
	29S10134A11	CONNECTOR, recept: for black, blue and brown leads
	14C68337A01	SHIELD, plastic: for R-G horiz line coil (L-1H)
	9C67580A03	SOCKET, molded: 12 pin

ELECTRICAL PARTS

MISCELLANEOUS ELECTRICAL PARTS

REF.	PART NO.	DESCRIPTION
E-1L thru E-6L	48C65837A02	DIODE, crystal
E-7L	48D67120A02	DIODE

REF.	PART NO.	DESCRIPTION	REF.	PART NO.	DESCRIPTION
	42S10122A01	CLIP, secures line cord to cab back		15D70134D01	COVER, ant terminal board:located on cab back (polystyrene)
	42S10152A02	CLIP, push on: spkr mtg		15V68642A20	COVER, cabinet back: complete; incls ant terminal board cover, rear tube cover & line cord (WU937 & WU938)
	42C67787A04	CLIP, spring: cab back mtg			
	42B70245A01	CLIP, spring catch: CRT bezel mtg			
	30S183A13	CORD, AC line		15V68642A08	COVER, cabinet back: complete; incls ant terminal board cover, rear tube cover & line cord (WU939)
	15D70134D01	COVER, ant terminal board: located on cab back (polystyrene)			
	15V68644A99	COVER, cabinet back: complete; incls ant terminal board cover, rear tube cover & line cord (WT901GWA)		15D68244A05	COVER, CRT: rear
				15P65175A41	DOOR, control: incls ctrl overlay; less cloth panel overlay
	15V68642A20	COVER, cabinet back: complete; incls ant terminal board cover, rear tube cover & line cord (WU911, 912, 913 & 914)		13P65175A43	ESCUTCHEON, control panel: UHF/ VHF; incls slide ctrl overlay & pre-set indicator jewel
				16E70396A03	GALLERY, cabinet (WU938GSA)
				16E70396A04	GALLERY, cabinet (WU938GUA)
	15E68244A05	COVER, CRT: rear		55C60850B72	GLIDE, dome: cab base front (WU937GUA & WU939GKA)
	15P65175A36	DOOR, control: incls ctrl overlay; less hinge spring clip			
	55C63659A28	GLIDE, dome: cab leg (WU912GWA)		13D69943A04	GRILLE, speaker: less mtg spring & cloth panel overlay
	13E69943A05	GRILLE, speaker: less mtg spring			
	42B70093A01	HINGE, spring clip: ctrl door (RH & LH)		42B70093A01	HINGE, spring clip: ctrl door (RH & LH)
	36D70324A04	KNOB, channel selector: UHF (outer)		36D70324A01	KNOB, channel selector: UHF (outer)
	36D70324A06	KNOB, channel selector: VHF		36D70324A03	KNOB, channel selector: VHF
	36D70325A04	KNOB, fine tune: UHF (inner)		36D70325A02	KNOB, fine tune: UHF (inner)
	36D69941A06	KNOB, on-off vol & VHF fine tune		36D69941A03	KNOB, on-off vol VHF fine tune
	36C69981A01	KNOB, slide: intensity & hue		36C69981A02	KNOB, slide: hue & intensity
	16B70115A01	LEG, cabinet (WU911GWA)		13C69934A15	OVERLAY, control (on-vol VHF fine tune)
	61A70527A01	LENS, color pre-set			
	13C69934A14	OVERLAY, control (on vol VHF fine tune)		13C69992A05	OVERLAY, control door (tone-vid peak-bright, etc.)
				13D69957A26	OVERLAY, control door: cloth panel (WU937GUA)
	13C69992A05	OVERLAY, control door (tone-vid peak-bright-etc.)			
	13C69993A03	OVERLAY, control panel: UHF/VHF slide ctrl; less pre-set lens		13D69957A28	OVERLAY, control door: cloth panel (WU938GSA, GUA)
	13C69934A13	OVERLAY, nameplate ("Quasar II")		13D69957A23	OVERLAY, control door: cloth panel (WU939GKA)
	64A67158A02	PLATE, metal: leg mtg (WU911GWA)			
	36D70491A02	PUSHBUTTON, color pre-set		55C60850B79	OVERLAY, decorative: cab base (WU937GUA)
	34D69996A02	SCALE, dial: UHF			
	34D69996A01	SCALE, dial: VHF		55C60850B73	OVERLAY, decorative (RS): cab base front (WU938GSA, GUA)
	3S136869	SCREW, tpg: 8-18 x 1-1/8 (phl flt blk ox) ctrl panel bezel mtg			
				55C60850B74	OVERLAY, decorative (LS): cab base front (WU938GSA, GUA)
	3D66303A18	SCREW, tpg: 8-15 x 3/4; CRT bezel to cab back			
	3D66303A19	SCREW, tpg: 8-15 x 13/16; interlock		55C60850B75	OVERLAY, decorative (RS): cab base side (WU938GSA, GUA)
	50D67337A01	SPEAKER, 4 x 6 PM 16Ω VC			
	41A70069A02	SPRING, sprk grille retainer		55C60850B76	OVERLAY, decorative (LS): cab base side (WU938GSA, GUA)

MODELS WU937GUA, WU938GSA,-GUA, WU939GKA

REF.	PART NO.	DESCRIPTION	REF.	PART NO.	DESCRIPTION
				55C60850B77	OVERLAY, decorative: cab base; front (R&L – WU939GKA)
	13P65175A42	BEZEL, control panel: incls ctrl esc & overlays; less ctrl door assembly & speaker grille		55C60850B78	OVERLAY, decorative: cab base; side (R&L – WU939GKA)
				13C69934A16	OVERLAY, nameplate (Quasar II)
	13E70276A01	BEZEL, CRT		13C69945A11	OVERLAY, slide control (intensity & hue)
	16E70395A01	CABINET, TV: console; trenwood (WU937GUA)			
	16E70396A01	CABINET, TV: console; maple (WU938GSA)		13D69957A27	OVERLAY, speaker grille: cloth panel (WU937GUA)
	16E70396A02	CABINET, TV:console; pine (WU938GUA)		13D69957A29	OVERLAY, speaker grille: cloth panel (WU938GSA, GUA)
	16E70397A01	CABINET, TV: console; oak (WU939GKA)		13D69957A24	OVERLAY, speaker grille: cloth panel (WU939GKA)
	55C63659A23	CASTER, less grip neck socket (WU937, WU938, & WU939)		36D70491A06	PUSHBUTTON: color pre-set
				34D69996A04	SCALE, dial: UHF
	42S10152A02	CLIP, push on: spkr mtg		34D69996A03	SCALE, dial: VHF
	42S10122A01	CLIP, secures line cord to cab back		3D66303A19	SCREW, tpg: 8-15 x 13/16; interlock
	42C67787A04	CLIP, spring: cab back mtg		3S136869	SCREW, tpg: 8-18 x 1-1/8 (phl flt blk ox) ctrl panel bezel mtg
	42B70245A01	CLIP, spring catch: CRT bezel mtg		55C63659A24	SOCKET, grip neck: for caster (WU937, WU938 & WU939)
	30S183A13	CORD, AC line		50D67337A01	SPEAKER, 4 x 6 PM 16Ω VC
				41A70069A02	SPRING, speaker grille retainer

CTV6

REF.	PART NO.	DESCRIPTION	REF.	PART NO.	DESCRIPTION
	UNIQUE ELECTRICAL & MECHANICAL REPLACEMENT PARTS FOR 16, E16, 18 & F18TS-929 CHASSIS. Also chassis code changes B-00 & later.		L-801	24V68639A28	COIL, CRT degausser: complete (16 & E16TS-929)
				24V68636A12	COIL, CRT degausser: complete (18 & F18TS-929)
	MISCELLANEOUS ELECTRICAL PARTS		PL-800	65S135685	LAMP, min incand: 6.3V #755; UHF (E16 & F18TS-929)
F-500	65S132920	FUSE, .5A 125V (B-00 & later)			

REF.	PART NO.	DESCRIPTION
E-4S	48C65837A02	CRYSTAL, diode
E-5S	48C65837A02	CRYSTAL, diode
COILS & CHOKES		
L-1S	24D68801A02	COMPENSATING: 450 uh
L-2S	24D68771A02	3.58Mc OSCILLATOR OUTPUT
L-3S	24D66772A14	CHOKE, 3.58Mc radiation supp
L-4S	24D68771A04	3.58Mc OSCILLATOR AMPLIFIER
L-5S	24D68771A03	3.58Mc OSCILLATOR TANK
L-6S	24D68801A02	COMPENSATING: 450 uh
TRANSISTORS		
Q-1S	48S134841	1ST COLOR IF (4841)
Q-2S	48S134841	2ND COLOR IF (4841)
Q-3S	48S134841	3.58Mc CW AMP (4841)
Q-4S	48S134733	ACC AMP (4733)(USE 48S134918 - A1L)
Q-5S	48S134910	1ST KILLER (P1C)
Q-6S	48S134842	2ND KILLER (4842)
Q-7S	48S134841	COLOR SYNC AMP (4841)
Q-8S	48S134933	3.58Mc OSCILLATOR A1V: use A3N
	48S137003	3.58Mc OSCILLATOR A3N
Q-10S	48S134842	COLOR SYNC GATE (4842)
Q-11S	48S134905	PHASE SPLITTER (A1H)
	48S134970	PHASE SPLITTER A2T in panel S-5A & S-6
Q-12S	48S134841	3.58Mc AMP (4841)
TRANSFORMERS		
T-1S	24G10282A05	1ST COLOR IF
T-2S	24G10282A01	2ND COLOR IF
T-3S	24G10282A03	COLOR, take-off
T-4S	24G10282A04	COLOR SYNC FILTER
MECHANICAL PARTS		
	29S10134A20	CONNECTOR, recept: PLAcir on panel (25 used)

POWER SUPPLY "J"

ELECTRICAL PARTS

CAPACITORS - NOTE: The capacitors in this list are recommended replacement types for the original equipment; all are ceramic disc type unless otherwise specified.

REF.	PART NO.	DESCRIPTION
C-2J	23C67753A06	500 mf/75V; 50 mf/350V; 50 mf/350V lytic
C-3J	23C67753A04	750 mf/150V; 500 mf/75V lytic
C-4J	23C67753A05	500 mf/150V; 500 mf/150V lytic
C-8J	21S129560	.001 mf 20% 1KV Z5F cer disc
C-9J	21S129560	.001 mf 20% 1KV Z5F cer disc
C-10J	21S129560	.001 mf 20% 1KV Z5F cer disc
MISCELLANEOUS ELECTRICAL PARTS		
E-1J	80C66390A20	CIRCUIT BREAKER
E-2J	40C68091A03	SWITCH, degaussing
	40D68091A04	SWITCH, degaussing: incls R-12J thermistor in PS-915-6-1/ PS-919-6-1 TS-915B-19-1/ TS-919B-17-1
E-3J	40C65513A10	SWITCH, defeat
E-3JA, B,C,D	48S191A05	RECTIFIER, silicon (4 used)
E-4J	48S191A06	RECTIFIER, silicon (USE 48S191A07)
E-9J	48S191A08	RECTIFIER, silicon TS-915B-12 TS-919B-10
COILS & CHOKES		
L-3J	25C65806A14	CHOKE, filter
L-4J	25C65806A14	CHOKE, filter
L-6J	25C65806A15	CHOKE, filter (USE 25D69148A01)
L-7J	25C65806A15	CHOKE, filter (USE 25D69148A01)
RESISTORS		
R-9J	17S135891	20 10% 20W WW
	17S135807	10 10% 15W WW in PS-2 (recomend replace with 10 10% 20W WW)
	17S135868	10 10% 20W WW in PS-915/919-6
R-10J	17S10130C70	750 10% 10W fxd mtl film
R-11J	17S135892	33 10% 10W WW

REF.	PART NO.	DESCRIPTION
R-12J	6C65884A09	THERMISTOR (NTC) 200Ω cold, 6Ω hot (solder in type)
	6P65147A70	THERMISTOR, complete: incls holder (TS-915B-13, TS-919B-11 - thermistor can be purchased separately. Part number listed below)
	6C65884A11	THERMISTOR (NTC) 200Ω cold, 6Ω hot (used with 6P65147A70 - used with 40D68091A04 switch in PS-915-6-1/PS-919-6-1/ TS-915B-19-1/TS-919B-19-1)
R-13J	17S135807	10 10% 15W WW in PS-2 (recomend replace with 10 10% 20W WW)
	17S135868	10 10% 20W WW in PS-915/919-6
TRANSFORMER		
T-1J	25D68580A01	POWER
MECHANICAL PARTS		
	42C66393A03	CLIP, retaining: WW resistor R-9J
	42S10137A04	CLIP, retaining: WW resistor R-11J
	15S10183A12	CONNECTOR, plug: 6 contacts; less contacts
	15S10183A14	CONNECTOR, plug: 10 contacts; less contacts
	39S10184A02	CONTACT (for 15S10183A12 and 15S10183A14)
	31C67063A03	PANEL, terminal cluster
	41C65987B14	SPRING, aquadag grounding
	41C68718A01	SPRING, back cover grounding (23TS-915)
	41A65351A06	SPRING, back cover grounding (23 & C23TS-919)

MISCELLANEOUS ELECTRICAL PARTS FOR TS-915 & TS-919 CHASSIS THAT ARE NOT ON PANELS

MISCELLANEOUS ELECTRICAL PARTS

REF.	PART NO.	DESCRIPTION
E-1C	80C68147A05	SPARK GAP
E-2C	80C68147A05	SPARK GAP
E-3C	80C68147A05	SPARK GAP
E-5A, B & C	65S135685	PILOT LAMP, 6.3V
E-7F	31B69649A01	STRIP, terminal: special, 8KV spark gap (on HV transf)
E-7J	51C67517A01	RES CAP (R-1J & C-5J)
E-10P	65S136193	LAMP, min. incand: 28ES (fine tune indicator)
E-11P	65S10081A06	LAMP, neon: 3AD (fine tune indicator - in some sets)
E-1R	48S134958	DIODE, silicon: D1H
E-4T	40V68617A74	SWITCH, FTL defeat: incls mtg brkt
COILS & CHOKES		
L-1J	24C68976A03	CHOKE, AC line filter
L-5J	24V68607A03	COIL, CRT degaussing: complete; incls leads and connectors
	24D68592A01	YOKE, deflection: 92° complete; incls pulg
	24P65146A78	DYN CONV: incls cores; less radial magnet & spring (red)
	24P65146A78	DYN CONV: incls cores; less radial magnet & spring (grn)
	24P65146A78	DYN CONV: incls cores; less radial magnet & spring (blue)
L-9B	24D66772A12	CHOKE, 7.5 uh (RF AGC line)
TRANSISTORS		
Q-5D	48S134920	AUDIO POWER AMP (A1N)
Q-1H	48S134900	VERT OUTPUT (A1C)
Q-1R	48S134936	H.V. REGULATOR (A1Y)
CONTROLS		
	40D66846A03	SWITCH, on/off; DPST 3 amp 125V AC (23TS-915)
R-1C	18D67502A12	G-2 GRN: 10 meg
R-3C	18D67502A11	G-2 BLUE: 10 meg
R-5C	18D68222A11	TINT: 4 meg (23TS-915)
	18D67502A16	TINT: 4 meg (23 & C23TS-919)
R-6C	18D67502A10	G-2 RED: 10 meg
R-7C	18D67502A13	FOCUS: 10 meg (23TS-915)

REF.	PART NO.	DESCRIPTION
R-15C	18D67502A20	FOCUS: 10 meg (23 & C23TS-919)
	18D67559A37	ABL: 1 meg
R-15D	18D68222A07	TONE: 50K (23TS-915)
	18D67559A41	TONE: 50K (23 & C23TS-919)
R-30D	18D68443A05	VOLUME: 20K (23TS-915)
	18D67562A04	VOLUME: incls on/off sw 20K DPST 125V AC 3A switch (23 & C23TS-919)
R-19E	18D68443A03	CONTRAST: 250 (23TS-915)
	18D67559A35	CONTRAST: 250 (23 & C23TS-919)
R-42E	18D68222A06	VERT HOLD: 100K (23TS-915)
	18D67559A33	VERT HOLD: 100K (23 & C23TS-919)
R-30L	18D68443A04	BRIGHTNESS: 750 (23TS-915)
	18D68222A09	BRIGHTNESS: 750 (23 & C23TS-919)
R-8S	18D68443A02	INTENSITY: 1.5K (23TS-915)
	18D68222A08	INTENSITY: 1.5K (23 & C23TS-919)
R-59S	18D68443A01	HUE: 800 (23TS-915)
	18D68222A10	HUE: 800 (23 & C23TS-919)

TRANSFORMERS

REF.	PART NO.	DESCRIPTION
T-1	24C68848A01	COIL, balum (antenna - on antenna board)
T-3D	25D68524A01	AUDIO OUTPUT
T-2F	24D68820A01	HORIZ OUTPUT & H.V.: complete; incls horiz pulse coil, Pri-Sec winding or horiz pulse coil may be purchased separately. Part numbers listed below.
	24D68820A03	HORIZ OUTPUT & H.V.: complete; incls horiz pulse coil, Pri-Sec winding may be purchased separately. Part numbers listed below (23TS-915B-07 & C, 23TS-919B-03)
	24D67565A06	PRI-SEC WINDING ONLY (part of T-2F)
	24D68819A01	COIL, horiz pulse (part of T-2F)
	25D68548A01	CRT FILAMENT (23TS-915 - USE 25D68548A02)
	25D68548A02	CRT FILAMENT (23 & C23TS-919)
	25D68548A05	CRT FILAMENT (TS-919-09 & later)

MECHANICAL PARTS

REF.	PART NO.	DESCRIPTION
	1P65146A96	ANTENNA BOARD, incls term strip: less balum coil (23TS-915)
	14D68726A02	ANTENNA BOARD, incls term strip & socket: less balum coil (23 & C23TS-919)
	7B69105A01	BRACKET, shaft support (noise inverter & AGC threshold - 23 & C23TS-919)
	7D67151B04	BRACKET, yoke: plastic; dynamic conv coil mtg
	75C66381A12	BUMPER, rubber: antenna board (23 & C23TS-919)
	42C65572A03	BUTTON, strap: plastic (degaussing coil mtg)
	30D67870A02	CABLE AND PLUG ASSEM: UHF TO VHF TUNER
	1V68606A84	CABLE AND PLUG: (audio take off coax - 23TS-915)
	1V68611A06	CABLE AND PLUG: (audio take off coax - 23 & C23TS-919)
	30P65147A13	CABLE AND PLUG CONVERTER (tuner to IF panel)
	30D67870A07	CABLE AND PLUG, shielded (antenna to VHF tuner - 23 & C23TS-919)
	1V68606A32	CABLE, shielded (hue control - 23TS-915)
	1V68610A96	CABLE, shielded (hue control - 23 & C23TS-919)
	30D67870A05	CABLE AND PLUG, shielded: incl clamp (antenna to VHF tuner - 23TS-915)
	30V68617A87	CABLE ASSEMBLY: gray; RF AG 3 feed back (C, E, & 23TS-919)
	30V68617A56	CABLE, shielded: incls connector, terminal & clamp (23TS-915B-00 FTL panel to FTL light - C23TS-919B-00 FTL panel to FTI light)
	30V68617A55	CABLE, shielded: incls connector, terminal, & 3 clamps; (C23TS-919B-00 - FTL panel to VHF tuner)
	30V68620A56	CABLE, shielded: FTL switch to VHF tuner (C23TS-919B-00)

REF.	PART NO.	DESCRIPTION
	30V68620A57	CABLE, shielded: FTL switch to UHF tuner (C23TS-919B-00)
	42B65954A03	CLAMP, plastic: UHF ant lead
	42S10283A01	CLAMP, plastic: yoke leads (on H.V. shield)
	42C66880A05	CLAMP, spring (T2F core mtg)
	42S10122A04	CLIP, hinge (HV cover and HV cover shield ext)
	42S10152A09	CLIP, metal (40Mc IF shield bottom cover - Panel B)
	42S10152A07	CLIP, metal (color board (S) mtg to chassis)
	42S10137A04	CLIP, metal (WW resistor R-2J mtg - on main chassis)
	42C67815A03	CLIP, plastic: 2nd anode lead dress (inside H.V. cage) and lead dress (located on tuner mtg bracket)
	42C67815A03	CLIP, plastic: H.V. tube filament & transf wire lead dress (inside H.V. cage)
	41C67955A02	CLIP, yoke retainer ring grounding
	42C66218A07	CLIP, spring nut (chassis & convergence brkt mtg)
	31C68421A01	CONNECTOR, panel: 5 pin; on chassis
	9C66133A06	CONNECTOR, plate cap (hi voltage rect - USE 9C66133A11)
	15S10183A04	CONNECTOR, plug: 3 contacts; less contacts (focus resistor)
	15S10183A12	CONNECTOR, plug: 6 contact; less contacts (filament transf - 23 & C23TS-919)
	15S10183A10	CONNECTOR, plug: 8 contacts; less contacts (def yoke)
	15S10183A22	CONNECTOR, plug: 10 contact; less contacts (H.V. cage)
	15S10183A16	CONNECTOR, recept: 1 contact; less contact (FTI)
	15S10183A03	CONNECTOR, recept: 3 contact; less contacts (focus resistor)
	15S10183A11	CONNECTOR, recept: 6 contact; less contacts (power supply & filament transf)
	15S10183A09	CONNECTOR, recept: 8 contact; less contacts (def yoke)
	15S10183A21	CONNECTOR, recept: 10 contact; less contacts (H.V.)
	15S10183A13	CONNECTOR, recept: 10 contact; less contacts (power supply)
	42C65864A34	CONNECTOR, 2nd anode lead: incl lead
	39S10184A08	CONTACT (for CRT leads-red-blue-grn)
	39S10184A02	CONTACT (for plugs 15S10183A10 - A12 & A22)
	39S10184A04	CONTACT (for plug 15S10183A04)
	39S10184A01	CONTACT (for recept 15S10183A03 - A09-A11-A13-A16-A21)
	39S10184A07	CONTACT (for video drive board ref # 6M 255V)
	1V68613A63	CORD, AC: interlock; incls brkt (on chassis)
	1V66588A06	CORD, dial: UHF (23TS-919)
	76C66210A06	CORE, iron: (T2F)
	38S10285A01	COVER, plastic: (covers AC plug on rear chassis panel)
4	44C67370A02	GEAR, driven: VHF fine tune (on chassis)
5	44C67370A06	GEAR & SHAFT, driver: VHF fine tune (on chassis 23TS-915, C, E, 23TS-919 - no FTL versions)
	5K752248	GROMMET, plastic: H.V. transf mtg (inside H.V. cage)
	5B68518A01	GROMMET, plastic: module panel mtg
	5A741077	GROMMET, strain relief: 2nd anode lead (on H.V. cage)
	14C68555A02	INSULATOR, armite: cable harness; 2 req (located on side of chassis)
	14D67783A21	INSULATOR, armite: focus control (23TS-915)
	14D67783A18	INSULATOR, armite: focus control (23 & C23TS-919)
	14D68555A07	INSULATOR, armite: lead dress (has 1/2" dia hole - 23 & C23TS-919)

REF.	PART NO.	DESCRIPTION
	14D68555A08	INSULATOR, armite: lead dress (has elongated hole - 23 & C23TS-919)
	14B68998B01	INSULATOR, armite: pushbutton UHF light shield
	14B65732A04	INSULATOR, connector: degausser leads
	14C68711A05	INSULATOR, cup: H.V. socket V1; incl wedge
	14D68067A07	INSULATOR, fibre: shields on-off control
	14B69139A01	INSULATOR, fibre: spring fuse resistor
	14D68555A05	INSULATOR, fibre: tint control mtg
	14D65732A06	INSULATOR, for spkr terminals
	14A562353	INSULATOR, mica: for transistor Q1H (USE 14A54381)
	14A63948A01	INSULATOR, mica: for transistor Q1R, Q5D
	14C68842A01	INSULATOR, plastic: (covers color board mtg clips)
	14D68555A04	INSULATOR, shield: pilot light (23 & C23TS-919)
	1V68613A53	LATCH, chassis stop: incls spring (on bottom chassis rail)
	59D68009A01	MAGNET, blue lateral & purity: incls mtg screw
	76D66816A06	MAGNET, radial: static convergence
	2S7051	NUT, 3/8-32 (control mtg)
	2S7981	NUT, push-on (FTI socket)
	2S10101A31	NUT, push-on: secures conv brkt assem to chassis
	2S10054A40	NUT, spring: filament transf mtg
	2B66148A02	NUT, T2F core mtg
	29K580544	PLUG, 72Ω antenna lead in
	28C66635A05	PLUG, AC (on chassis)
	28C67679A03	PLUG, molded: 12 pin; convergence coil assembly
	52B68512B01	POINTER, UHF (23TS-919)
	1V66588A07	PULLEY, UHF on mtg brkt (23TS-919)
	9C66402A04	RECEPTACLE, degausser coil lead
	29A739899	RECEPTACLE, spkr leads (23TS-915)
	9C66402A05	RECEPTACLE, spkr leads (23 & C23TS-919)
6	42C67369A01	RETAINER, plastic: gear; VHF fine tune (on chassis)
	42D69225A01	RETAINER, ring: metal; yoke mtg
	28S10114A10	RIVET, plastic: HV cage (23 & C23TS-919)
	5S10281A01	RIVET, plastic: (hinge for video drive & output brkt)
	5S7836	RIVET, snap in (secures 14D68067A07 on-off insulator)
	34D68261C01	SCALE, dial VHF
	34P65147A01	SCALE, UHF incls brkt (plastic - 23 & E23TS-919)
	3S135356	SCREW, tpg: 8 x 1-3/4 (yoke ring mtg)
	3D66303A06	SCREW, tpg: 10-12 x 1-1/4 (chassis mtg)
	3D66303A15	SCREW, tpg: 10-16 x 1/2 (tuner mtg brkt)
	47P65147A80	SHAFT, VHF fine tune: incls drive gear, retainer gear & bobbin to activate FTL switch (23TS-915B-00)
	47P65147A81	SHAFT, VHF fine tune: incls drive gear, retainer gear & bobbin to activate FTL switch (C23TS-919B-00)
	47B69107A01	SHAFT, plastic: for noise invertor & AGC threshold (23 & C23TS-919)
	14D68067A06	SHIELD, fibre: pilot light (23TS-919)
	26B65584A05	SHIELD, metal: VHF pilot light
	43B65441A04	SLEEVE, rubber: located on CRT mtg bracket
	9D66770A09	SOCKET, 12 pin: V1 (USE 9D66770A05)
	1V68606A46	SOCKET, CRT: complete; incls leads, connectors & resistors (23TS-915)
	9V68619A27	SOCKET, CRT: complete; incls leads, connectors & resistors (23TS-915B-00)
	1V68610A91	SOCKET, CRT: complete; incls leads, connectors & resistors (23 & C23TS-919)
	9V68619A30	SOCKET, CRT: complete; incls leads, connectors & resistors (C, E, 23TS-919B-00)

REF.	PART NO.	DESCRIPTION
	1V68609A50	SOCKET, single FTI: incls term strip (in some sets)
	9P65147A31	SOCKET, DUAL: FTI; incls term strip (in some sets)
	9V68617A65	SOCKET, single: FTL incls term strip (23TS-915B-00)
	9S10143A16	SOCKET, pilot light & brkt (23TS-915)
	9S10143A15	SOCKET, pilot light & brkt (23 & C23TS-919)
	9C63825A01	SOCKET, transistor: Q1H
	9C67532A02	SOCKET, transistor: Q1R, Q5D
	9B63949A01	SOCKET, transistor: Q5D (23 & C23TS-919 only)
	41A65351A05	SPRING, back cover ground (located on chassis)
	41C65987A09	SPRING, bezel grounding
	41A69140A01	SPRING, fuse resistor
	41A67281A02	SPRING, magnetic shield retainer
	41C66795A02	SPRING, retaining: radial magnets on convergence assem
	41A69106A01	SPRING, secures shaft of noise invertor & AGC threshold (23 & C23TS-919)
	41C65597A10	SPRING, tension: dial cord (23TS-919)
	42C65572A05	STRAP, adjustable type (for mtg degausser coil)
	42C65572A04	STRAP, button type: less buttons (for mtg degausser coil)
	42D67027A03	STRAP, CRT mtg
	4S2638	WASHER, lock: #10 int-ext (tuner mtg brkt)

CHS-62299

(ALSO SEE UNIQUE PARTS TO STEREO PORTION)

ELECTRICAL PARTS

MISCELLANEOUS ELECTRICAL PARTS

REF.	PART NO.	DESCRIPTION
	65S135846	FUSE: 3A-250V
E-24	48S134790	RECTIFIER, silicon: M4790
E-25	48S134790	RECTIFIER, silicon: M4790
E-29	28D62887A03	PLUG, 9-conn: less contacts; transf

TRANSFORMERS

REF.	PART NO.	DESCRIPTION
T-1	25D63502A01	DRIVER
T-2	25D63502A01	DRIVER
T-3	25D60061B01	POWER

TRANSISTORS

REF.	PART NO.	DESCRIPTION
Q-1	48S134733	M4733
Q-2	48S134732	M4732
Q-3	48S134903	A1F
Q-4	48S134766	SP-1483
Q-5	48S134766	SP-1483
Q-6	48S134733	M4733
Q-7	48S134732	M4732
Q-8	48S134903	A1F
Q-9	48S134766	SP-1483
Q-10	48S134766	SP-1483

MECHANICAL PARTS

REF.	PART NO.	DESCRIPTION
	42A61128A01	CLAMP, FM antenna
	39D62885A01	CONTACT, plug: use with 28D62887A03 & 28D62887A01 trans plug
	39D62885A02	CONTACT, recept: use with 15S10183A07 & 15S10183A17 recept
	39C66357A01	CONTACT, recept: use with 9A60402A01 B+ recept
	30S189A02	CORD, AC line
	14A543810	INSULATOR, trans
	29A64778A02	LUG, recept: use with 15S10183A07 recept
	2S7051	NUT, pal: 3/8-32 (cont & earphone recept mtg)
	28B63547A01	PLUG, line cord: interlock
	9C60671A09	RECEPTACLE, earphone
	9B60410B01	RECEPTACLE, fuse

REF.	PART NO.	DESCRIPTION	REF.	PART NO.	DESCRIPTION
	15S10183A17	RECEPTACLE, 2-conn; less contacts: bass switch		30C60053B01	CABLE, power: 4-conn; incl plug & recept; tuner to chgr
	9C60402A01	RECEPTACLE, 3-conn; less contacts. (tuner B+)		30D63525A04	CABLE, twin phono: gray 33" (to rec chgr - USE 30C63525A03)
	15S10183A07	RECEPTACLE, 9-conn; less contacts: (chassis to power transf)		39S10184A06	CONTACT, plug: multiplex board
	3S132845	SCREW, tpg: #6 x 1/2 (trans mtg)		39C66357A01	CONTACT, plug: use w/28A60398A01 B+ plug
	9B542339	SOCKET, trans		39S10184A05	CONTACT, recept: FM ant & multiplex lead

CHS-62298

(ALSO SEE UNIQUE PARTS TO STEREO PORTION)

REF.	PART NO.	DESCRIPTION	REF.	PART NO.	DESCRIPTION
		ELECTRICAL PARTS		58B63996A01	COUPLER, AM var cap to FM tuner
				5S135606	EYELET, mtg: to secure AM & FM tuner
MISCELLANEOUS ELECTRICAL PARTS				32B60851B01	GASKET, spacer: escut
E-3	48B62334A01	DIODE, crystal		5S10115A07	GROMMET, rubber: to secure AM & FM tuner
E-4	48B62334A01	DIODE, crystal		9C66402A05	LUG, recept: AM ant
E-5	48B62334A01	DIODE, crystal		2S10054A21	NUT, clip-on: 6-32 (to secure function sw)
E-6	48B62334A01	DIODE, crystal		2S10080A04	NUT, hex: 3/8-32 (to secure tuning shaft bushing)
E-7	48B62334A01	DIODE, crystal		2S10101A15	NUT, push-on: to secure multiplex light rod
E-8	48B62334A01	DIODE, crystal		52B60611B01	POINTER, dial: incl slider
E-9	48B62334A01	DIODE, crystal		1V63035A99	PULLEY & BUSHING ASSEM: tuning
E-13	65S134388	BULB, multiplex: 10V-100EFC		38B63393A03	PUSHBUTTON: function sw
E-15	65S125595	BULB, dial: .15A-6 to 8V (AM-FM)		9D645314	RECEPTACLE, twin phono: blk & wht (audio output)
E-16	65S125595	BULB, pilot: .15A-6 to 8V (comp lite)		47C63516A01	ROD, light cond: multiplex
E-17	48S134851	DIODE, Zener: M4851		3S114254	SCREW, set: 4-40 x 3/8 (to secure var cap coupler & pulley & bushing assem)
E-19	48B62334A01	DIODE, crystal		1V63038A02	SLIDER BRACKET ASSEM: incl pointer; pulley
E-20	48B62334A01	DIODE, crystal		9S10143A11	SOCKET, dial light: AM-FM
E-21	28A60398A01	PLUG, 3-conn: less contacts (B+)		41A471681	SPRING, tension: dial cord
				1V63060A71	TUNING SHAFT & PULLEY ASSEM
COILS & CHOKES				4K600617	WASHER, "C": tuning shaft

REF.	PART NO.	DESCRIPTION
L-4	24C60581B01	ANTENNA, ferrite core: AM
L-5	24C63713A01	OSCILLATOR: AM
L-6	24C63463A01	MULTIPLEX: 19Kc
L-7	24C63463A01	STORECAST: 67Kc
L-8	24C63463A01	STORECAST: 67Kc
L-9	24C61620A04	CHOKE, B+
L-10	24C60093B01	PEAKING, AM antenna

CONTROL

REF.	PART NO.	DESCRIPTION
R-67	18B63499A01	GAIN: 19Kc

SWITCHES

REF.	PART NO.	DESCRIPTION
E-18	1V63052A37	SWITCH & REGULATOR ASSEM: incls switch, board & all components
	40D63177A01	SWITCH, function: less pushbuttons (part of E-18)
	40C63331A02	SWITCH, AC only (Part of E-18)

TRANSFORMERS

REF.	PART NO.	DESCRIPTION
T-2	24C63708A01	FM IF: 10.7Mc
T-3	24C63456A01	FM IF: 10.7Mc
T-4	24C63456A01	FM IF: 10.7Mc
T-5	24C60097B01	RATIO DETECTOR
T-6	24C63549A06	AM IF: 455Kc
T-7	24C63549A06	AM IF: 455Kc
T-8	24C60098B02	AM DETECTOR: 455Kc
T-9	24C63463A02	DOUBLER: 38Kc

TRANSISTORS

REF.	PART NO.	DESCRIPTION
Q-4	48S134857	M4857
Q-5	48S134857	M4857
Q-6	48S134857	M4857
Q-7	48S134857	M4857
Q-8	48S134960	A2L
Q-9	48S134826	M4826
Q-10	48S134906	A1J
Q-11	48S134906	A1J
Q-12	48S134768	M4768
Q-13	48S134768	M4768

REF.	PART NO.	DESCRIPTION
		MECHANICAL PARTS
	64C60612B01	BACKGROUND, dial
	1V63057A17	BOARD, plated chassis: incl contacts; multiplex
	84C60152B04	BOARD, plated: function sw
	43A63336A01	BUSHING, tuning shaft

CABINET PARTS

MODELS WL812CW, WL813CK, WL814CM, CW, WL815CS, WL816CS, WL817CC, WD831DU, WD832DW, WD833DU AND WD834DU

REF.	PART NO.	DESCRIPTION
	13P65146A19	BEZEL, control panel; incls: overlays, esc., nameplate, dial light indicator (WL813, WL815, WL816, WL817, WD831, WD832, WD833, WD834 series)
	13P65147A67	BEZEL, control panel; incls: overlays, esc., nameplate, dial light indicator (WL812, WL813, WL814, WL815, WL816, WL817-1 series - WD831, WD832, WD833, WD834-1 series)
	13E68570A01	BEZEL, CRT
	38P65146A21	BUTTON, on-off; incls push on-off insert
	13P65146A20	BUTTON, overlay; incls "M" insert
	16E68740A01	CABINET, console: walnut (WL812CW)
	16E68741A01	CABINET, console: oak (WL813CK)
	16E68750A01	CABINET, console: mah (WL814CM)
	16E68750A02	CABINET, console: walnut (WL814CW)
	16E68750A03	CABINET, console: walnut (WL815CW)
	16E68754A01	CABINET, console: maple (WL816CS)
	16E68754A02	CABINET, console: cherry (WL817CC)
	16E69005A01	CABINET, console: hespania (WD831DU)
	16E69006A01	CABINET, console: pecan (WD832DW)
	16E69007A01	CABINET, console: oak (WD833DU)
	16E69009A01	CABINET, console: pecan (WD834DU)
	55C63659A23	CASTER, cabinet (WL816CS, WD832DW, WD833DU)

MODELS WD825DU, WD826DD, WD827DU, WL850DW, WL851DW, WL853DK, WL855DC AND WL856DS

REF.	PART NO.	DESCRIPTION
	13P65147A18	BEZEL, CRT: incls door & overlays (WD825DU, WD826DD, WD827DU)
	13P65147A69	BEZEL, CRT: incls door & overlays (WD825DU-1, WD826DD-1 WD827DU-1)
	13P65147A17	BEZEL, CRT: incls door & overlays (WL850DW, WL851DW, WL853DK, WL855DC, WL856DS)
	75C66381A10	BUMPER, rubber: spkr panel (on bezel)
	16E68817A01	CABINET, console: pecan (WD825DU)
	16E68844A01	CABINET, console: mah (WD826DD)
	16E68845A01	CABINET, console: umber (WD827DU)

REF.	PART NO.	DESCRIPTION	REF.	PART NO.	DESCRIPTION
	16E68895A01	CABINET, console: wal (WL850DW)		13P65147A18	BEZEL, CRT: incls door & overlays (MK843DW, MK844DK, MK845DK)
	16E68951A01	CABINET, console: wal (WL851DW)		13P65147A69	BEZEL, CRT: incls door & overlays (MK843DW-1, MK844DK-1, MK845DK-1)
	16E68953A01	CABINET, console: oak (WL853DK)		75C66381A10	BUMPER, rubber: spkr panel (on bezel)
	16E68987A01	CABINET, console: cherry (WL855DC)		16E60816B01	CABINET, console: walnut (MK840DW)
	16E68954A01	CABINET, console: maple (WL856DS)		16E60818B01	CABINET, console: oak (MK841DK)
	55C63659A23	CASTER, (WD825DU, WD826DD)		16E60817B01	CABINET, console: walnut (MK843DW)
MODELS MK840DW, MK841DK, MK843DW, MK844DK, AND MK845DK				16E60813B01	CABINET, console: oak (MK844DK)
	13P65147A17	BEZEL, CRT: incls door & overlays (MK840DW, MK841DK)		16E60813B02	CABINET, console: oak (MK845DK)

REPLACEMENT PARTS FOR CHASSIS "C" 23TS- 915

REF.	PART NO.	DESCRIPTION	REF.	PART NO.	DESCRIPTION
		ELECTRICAL PARTS			**COILS & CHOKES**
L-5J	24V68628A07	COIL, CRT degaussing: complete; incls leads & connector	L-3J	25C65806A14	CHOKE, filter
			L-4J	25C65806A14	CHOKE, filter
T-2J	*25D68548A07	CRT FILAMENT TRANSFORMER	L-6J	25D69148A01	CHOKE, filter
			L-7J	25D69148A01	CHOKE, filter
		H.V. TRANSFORMER MODULE	L-12 (Z)	24C68976A02	CHOKE (in RP/S915-2 & later)
		ELECTRICAL PARTS			**TRANSFORMER**
T-2F	24V68628A89	H.V. TRANSFORMER: complete; incls "L" mtg brkt, focus resistor, solid state rectifier & all other associated parts (listed below are parts that may be purchased separately from this assembly)	T-1J	25G10365A01	POWER
					MECHANICAL PARTS
	24D69776B01	H.V. TRANSFORMER: only; incls pri-sec winding and pulse coil: less solid state rectifier & & insulator cup		42C66393A04	CLIP, WW resistor retainer: (R9J & R13J)
				42S10137A04	CLIP, WW resistor retainer: (R11J)
	24D67565A10	PRI-SEC WINDING, only		42S10137A06	CLIP, WW resistor retainer: (R15J & R16J)
	24D68819A01	COIL, horiz pulse		31C68421A01	CONNECTOR, panel: 5 pin; on regulator panel mtg brkt
	48D69723A01	SOLID STATE RECTIFIER		15S10183A12	CONNECTOR, plug: 6 contact; less contacts
E-7F	31B69649A01	STRIP, terminal: special; 8KV spark gap		15S10183A14	CONNECTOR, plug: 10 contact; less contacts
R-26F	6D68703A01	FOCUS RESISTOR		29S10134A30	CONNECTOR, spade lug: (R15J & R16J)
R-30F	17S750273	1.0 10% 5W WW resistor (USE 17S561979)		39S10184A02	CONTACT (for 15S10183A12 & 15S10183A14)
		MECHANICAL PARTS		5C68518C01	GROMMET, plastic: regulator panel mtg (2 used)
	42D65864A45	CONNECTOR, 2nd anode lead: incls lead		31C67063A03	PANEL, terminal cluster
	9D66133A06	CONNECTOR, plate cap: solid state rectifier (USE 9D66133A11)			**REGULATOR PANEL "Z"**
	39S10184A02	CONTACT, plug (for 15S10183A22)			**ELECTRICAL PARTS**
	15S10183A22	CONNECTOR, plug: 10 contact; less contacts			**MISCELLANEOUS ELECTRICAL PARTS**
	39S10184A01	CONTACT, recept (for 15S10183A03)	E-1Z	48V68629A40	THYRISTOR: W1F; incls heat sink
	15S10183A03	CONNECTOR, recept: 3 contact; less contacts (focus resistor)	E-2Z	48S137063	DIODE, silicon: D2E
	5K752248	GROMMET, tpg screw: plastic; H.V. transformer mtg	E-3Z	48S137065	DIODE, silicon: D2F
	14C68711A05	INSULATOR, cup: rectifier socket; incl wedge	E-4Z	48G10346A01	DIODE, (USE 48D67120A11)
	3S124796	SCREW, tpg: 6 x 1/2; rectifier socket mtg	E-5Z	48G10346A01	DIODE, (USE 48D67120A11)
	42A69711A01	SOCKET, solid state rectifier mtg			**TRANSISTORS**
		POWER SUPPLY "J" (REGULATOR)	Q-1Z	48S134910	REG AMP: P1C
		ELECTRICAL PARTS	Q-2Z	48S134842	REG SYNC: M4842 (USE 48S134992)
		MISCELLANEOUS ELECTRICAL PARTS			**CONTROLS**
E-1J	80C6390A20	CIRCUIT BREAKER	R-5Z	18D66401A24	REGULATOR: 10K
E-3J	40C65513A10	SWITCH, defeat (USE 40C65513B10)			**TRANSFORMERS**
E-3J A,B, C & D	48S191A05	RECTIFIER, silicon (4 used - USE 48S191A07)	T-3Z	*24D69724A01	TRIGGER
			T-4Z	*25D68865A02	SYNC
E-4J	48S191A07	RECTIFIER, silicon			
E-9J	48S191A08	RECTIFIER, silicon			**MECHANICAL PARTS**
E-11J	6C69717A01	RESISTOR, PTC: degausser		29S10134A29	CONNECTOR, receptacle: Placir chassis mtg (5 used)
				29S10134A32	LUG, terminal: (ref 6Z)

REF.	PART NO.	DESCRIPTION	REF.	PART NO.	DESCRIPTION
		CABINET PARTS		36D68463A01	KNOB, control: tint-vertical-tone
				36D68218A25	KNOB, fine tune
MODEL WD841EN				36D68458A02	KNOB, slide: chrome; hue-intensity-volume
	13P65149A80	BEZEL, control panel: incls overlays, escutcheon, nameplate, dial light indicator lens; less speaker grille		36D68458B03	KNOB, slide: contrast, bright
				36D68217A35	KNOB, UHF channel selector
				36D68217A29	KNOB, VHF channel selector
	13E68570A01	BEZEL, CRT		16E61245B10	LID, cabinet LH; AM/FM compartment
	38P65146A21	BUTTON, on-off: incls overlay (push on-off)		16E61245B11	LID, cabinet: RH; record changer compartment
	16F69380A02	CABINET, console: pecan		33C68940A01	NAMEPLATE (cabinet by Drexel)
	55C63659A23	CASTER		33C68104B08	NAMEPLATE (Motorola "Quasar")
	42S10122A01	CLIP, AC cord retainer: secures line cord to cab back		2S10080A09	NUT, 6-32: speaker grille mtg
	42S10258A02	CLIP, push-on: secures nameplate (cab by Drexel) to cabinet		13P65149A81	OVERLAY, control panel: slide switch; incls overlay "M"
	42S10152A02	CLIP, push-on: spkr mtg		55C60850B37	PULL, door
	42C67787A01	CLIP, spring: cab back cover mtg		34D68464A01	SCALE, dial: VHF
	30S183A13	CORD, AC line		3S134276	SCREW, 6 x 3/8: 45 RPM spindle clip mtg (2 req'd)
	15E68757A10	COVER, cabinet back		55C63659A24	SOCKET, grip neck: for casters
	15E68244A03	COVER, CRT: rear		50D67952B03	SPEAKER, 6" PM: 8Ω TV
	13P65149A82	ESCUTCHEON, control panel: center; incls nameplate		41C65597B11	SPRING, on-off button return
	13E68777A26	GRILLE, speaker: incls grille cloth		55C60978A10	SUPPORT, lid: self-balancing
	55C63659A24	SOCKET, grip neck: for casters			
	36D68463A01	KNOB, control (tint-vertical-tone)	**MODEL WU870 & WU871EW**		
	36D68218A25	KNOB, fine tune		13P65172A54	BEZEL, control panel: incls slide control overlay, "M" insert, dial light indicator lens; less speaker grille & UHF/VHF fine tune escutcheon
	36D68458A02	KNOB, slide: chrome (hue-intensity-volume)			
	36D68458B03	KNOB, slide: olive green (contrast-brightness)		13E68570A04	BEZEL, CRT
	36B68217A35	KNOB, UHF channel selector		75C69738A01	BUMPER, cab foot: blk (plastic - WU870EW)
	36D68217A29	KNOB, VHF channel selector		38P65146A21	BUTTON, on-off: incls push on off insert
	33C68940A01	NAMEPLATE (cabinet by Drexel)		16E69737A01	CABINET, console: walnut; less legs & stretcher (WU870EW)
	33C68104B08	NAMEPLATE (Motorola-"Quasar")		16E69742A01	CABINET, console: walnut; less legs & stretcher (WU871EW)
	2S10080A09	NUT, 6-32: speaker grille mtg		42S10152A02	CLIP, push-on (speaker mtg)
	13P65149A81	OVERLAY, control panel: slide switch; incls overlay "M"		42S10122A01	CLIP (secures AC cord to cab back)
	34D68464A01	SCALE, dial: VHF		42C67787A01	CLIP, spring: back cover mtg to cabinet
	50D67952B03	SPEAKER, 6" PM: 8Ω		30S183A13	CORD, AC line
	41C65597B11	SPRING, on-off button return		15E68757A15	COVER, cabinet back
				15E68244A03	COVER, CRT: rear
MODEL MD850EN				13P65170A52	ESCUTCHEON UHF/VHF & FINE TUNE CONTROL: incls Motorola "Quasar" UHF overlay
	13P65149A80	BEZEL, control panel: incls overlays, escutcheon, nameplate, dial light indicator lens; less speaker grille			
				55C63659A28	GLIDE, dome: cab leg
	13E68570A01	BEZEL, CRT		13E68777A25	GRILLE, speaker: incls grille cloth (WU871EW)
	38P65146A21	BUTTON, on-off: incls overlay (push on-off)		13D69710B01	GRILLE, speaker (WU870EW)
	16E61245B07	CABINET, stereo TV console: hispania		14D68555A06	INSULATOR, lens retainer
	55B60850B38	CASTER		36D68218A25	KNOB, fine tune
	42S10122A01	CLIP, AC line cord retainer: secures line cord to cabinet back		36C68458B03	KNOB, slide (contrast & brightness)
	42S10258A02	CLIP, push-on: secures nameplate (cabinet by Drexel) to cabinet		36C68458A02	KNOB, slide (hue, intensity, & volume)
	42S10152A02	CLIP, push-on: speaker mtg		36C68463A01	KNOB (tint, vertical, tone)
	42B62856A05	CLIP, spindle: 45 RPM		36D68217A30	KNOB, UHF channel selector
	42C67787A01	CLIP, spring: cab back cover mtg		36D68217A29	KNOB, VHF channel selector
	35D60005B49	CLOTH, grille		16B69622A01	LEG, cabinet (WU870EW)
	30C67741A01	CORD, AC line with recept outlet		16B69740A01	LEG, cabinet (WU871EW)
	15E61259B01	COVER, cabinet back: AM/FM section		61B68582B04	LENS, indicator light
	15E61259B02	COVER, cabinet back: record changer section		2S7981	NUT, push-on (indicator light lens mtg)
	15E68757A09	COVER, cabinet back: TV section		13P65170A54	OVERLAY, slide control: incls "M" insert
	15E68244A03	COVER, CRT: rear		64A67158A02	PLATE, leg mtg
	16E61245B08	DOOR, cabinet: LH		34D68464A01	SCALE, dial: VHF
	16E61245B09	DOOR, cabinet: RH		50D67952B03	SPEAKER, 6" PM: 8Ω
	13P65149A82	ESCUTCHEON, control panel: center; incls nameplate		41C65597A14	SPRING, comp: on-off button
	13E68777A23	GRILLE, speaker: incls grille cloth		16C69771A01	STRETCHER, cab leg (WU870EW)
	55C60850B05	HINGE, butt: cabinet lid		16C69741A01	STRETCHER, cab leg (WU871EW)

C25TS-915

REF.	PART NO.	DESCRIPTION	REF.	PART NO.	DESCRIPTION
MECHANICAL PARTS				15S10183A15	CONNECTOR, plug: 1 contact; less contact AFT
	29S10134A29	CONNECTOR, recept: module panel mtg		39S10184A02	CONTACT, plug: for connector 15 S10183A15)

REF.	PART NO.	DESCRIPTION
	9C67349A07	RECEPT, phono: UHF input (on tuner brkt)
	9C67349B02	RECEPT, phono: IF output (on module panel)

IF AUDIO PANEL "D" (CODED D-12 & LATER)

REF.	PART NO.	DESCRIPTION
	1V65534A84	IC AUDIO PANEL "D": complete
INTEGRATED CIRUITS		
IC-2	51D70177A02	INTEGRATED CIRCUIT
COILS		
L-1	24D68517A20	QUADRATURE
MECHANICAL PARTS		
	29S10134A29	CONNECTOR, recept: module panel mtg
	9C67349B03	RECEPTACLE, phono: 4.5MHz audio IF input

AFC PANEL "K"

REF.	PART NO.	DESCRIPTION
	1U65534A97	AFC PANEL: complete
MISCELLANEOUS ELECTRICAL PARTS		
E-1K	48C65837A02	DIODE, crystal
E-2K	48C65837A02	DIODE, crystal
COILS & CHOKES		
L-2K	24D66772A12	CHOKE, resonant
TRANSISTORS		
Q-1K	48S134937	AFC AMP, A1Z
Q-2K	48S134937	AFC OUTPUT, A1Z
TRANSFORMERS		
T-1K	24D68517A11	DISCRIMINATOR
MECHANICAL PARTS		
	15S10183A15	CONNECTOR, plug: 1 contact; less contact (AFC output)
	39S10184A02	CONTACT, plug: for 15S10183A15 connector

REMOTE CONTROL PANEL "Y" (CODED Y-7)

REF.	PART NO.	DESCRIPTION
	1U68674A02	REMOTE CONTROL PANEL "Y": complete
ELECTRICAL PARTS		
MISCELLANEOUS ELECTRICAL PARTS		
E-1Y	48D67120A06	DIODE (USE 48P65145A74)
E-2Y	48D67120A06	DIODE (USE 48P65145A74)
E-3Y	48S137074	DIODE: D2J
E-4Y	48S137074	DIODE: D2J
E-5Y	48D67120A06	DIODE
E-6Y	48D67120A06	DIODE
E-7Y	48S137074	DIODE: D2J
E-8Y	48S137074	DIODE: D2J
E-9Y	48D67120A06	DIODE (USE 48P65145A74)
E-10Y	48D67120A06	DIODE (USE 48P65145A74)
E-11Y	48S137074	DIODE: D2J
E-12Y	48S137074	DIODE: D2J
E-19Y	51D69344A01	MODULE, audio memory
E-20Y	51D69344A01	MODULE, hue memory
E-21Y	51D69344A01	MODULE, intensity memory
COILS & CHOKES		
L-1Y	24D66772A07	CHOKE, RF suppressor
L-2Y	24D68801A01	COMPENSATING: 15 uh
TRANSISTORS		
Q-1Y	48S134942	FUNCTION OUTPUT: A2B
Q-2Y	48S137014	1ST AUDIO AMP: A4A
Q-3Y	48S137014	2ND AUDIO AMP: A4A
Q-4Y	48S134841	1ST HUE AMP: M4841
Q-5Y	48S134841	2ND HUE AMP: M4841
Q-6Y	48S134841	INTENSITY AMP: M4841
CONTROLS		
R-62Y	18D66401A26	AUDIO CUT IN: 100K
TRANSFORMERS		
T-1Y	24D69051A03	VOLUME DOWN
T-2Y	24D69051A03	VOLUME UP
T-3Y	24D69051A03	HUE TO RED
T-4Y	24D69051A03	HUE TO GREEN
T-5Y	24D69051A03	INTENSITY DOWN
T-6Y	24D69051A03	INTENSITY UP
CONTROL VOLTAGE SWITCH PANEL (PART OF COMPLETE "Y" PANEL)		
TRANSISTOR		
Q-7	48S134841	HUE OUTPUT: M4841
SWITCH		
SW-1Y	*40D69977A02	SLIDE SWITCH
MECHANICAL PARTS		
	29S10134A29	CONNECTOR, recept: module panel mtg
	15S10183A16	CONNECTOR, recept: 1 contact; less contact (hue)
	39S10184A01	CONTACT, recept: for connector 15S10183A16

REGULATOR PANEL "Z" – FOR REGULATED AC POWER SUPPLY ONLY

REF.	PART NO.	DESCRIPTION
MISCELLANEOUS ELECTRICAL PARTS		
E-1Z	48V68629A40	THYRISTOR: W1F; incls heat sink
E-2Z	48S137063	DIODE, silicon: D2E
E-3Z	48S137065	DIODE, silicon: D2F
E-4Z	48G10346A01	DIODE (USE 48D67120A11)
E-5Z	48G10346A01	DIODE (USE 48D67120A11)
TRANSISTORS		
Q-1Z	48S134910	REG AMP: P1C
Q-2Z	48S134842	REG SYNC: M4842 (USE 48S134992)
CONTROL		
R-5Z	18D66401A24	REGULATOR: 10K
	18D66401A37	REGULATOR: 10K (in panel Z2 & later)
TRANSFORMERS		
T-3Z	24D69724A01	TRIGGER
T-4Z	25D68865A02	SYNC
MECHANICAL PARTS		
	29S10134A29	CONNECTOR, recept: panel mtg (5 used)
	29S10134A32	LUG, terminal (ref #6Z)

REGULATED POWER SUPPLY "J"

REF.	PART NO.	DESCRIPTION
MISCELLANEOUS ELECTRICAL PARTS		
E-1J	80C66390A20	CIRCUIT BREAKER
E-3J A,B,C, &D	48S191A05	RECTIFIER, silicon: 4 used (USE 48S191A07)
E-4J	48S191A07	RECTIFIER, silicon
E-8J	40C65513B10	SWITCH, defeat
E-9J	48S191A08	RECTIFIER, silicon
E-11J	6C69717A01	RESISTOR, PTC: degausser
COILS & CHOKES		
L-3J	25C65806A14	CHOKE, filter
L-4J	25C65806A14	CHOKE, filter
L-6J	25D69148A01	CHOKE, filter
L-7J	25D69148A01	CHOKE, filter
L-12Z	24C68976A02	CHOKE
TRANSFORMERS		
T-1J	25G10365A01	POWER
MECHANICAL PARTS		
	42C66393A04	CLIP, WW resistor retainer (R9J & R13J)
	42S10137A04	CLIP, WW resistor retainer (R11J)

REF.	PART NO.	DESCRIPTION
	42S10137A06	CLIP, WW resistor retainer (R15Z & R16Z)
	31C68421B02	CONNECTOR, panel: 5 pin; on regulator panel mtg brkt
	15S10183A12	CONNECTOR, plug: 6 contact; less contacts
	15S10183A14	CONNECTOR, plug: 10 contact; less contacts
	29S10134A30	CONNECTOR, spade lug (R15Z & R16Z)
	39S10184A02	CONTACT, plug: for 6 pin & 10 pin connector plug
	31C67063A03	PANEL, terminal cluster

HV TRANSFORMER MODULE ELECTRICAL & MECHANICAL PARTS

REF.	PART NO.	DESCRIPTION
T-2F(B)	24P65174A08	HV TRANSFORMER, incls pri-sec winding, pulse coil & 10 pin plug assembly: less mtg brkt, solid state rectifier & insulator cup assembly
	48D69723B01	RECTIFIER, solid state
E-7F	31B69649A01	STRIP, terminal: special; 8Kv spark gap
R-26F	6D68703A01	FOCUS RESISTOR
R-30F	17S750273	1.0 10% 5W WW resistor (USE 17S132815)

MECHANICAL PARTS

REF.	PART NO.	DESCRIPTION
	42S10122A04	CLIP, hinge: HV cover
	42C67815B03	CLIP, wire: plastic; lead dress
	42D65864A57	CONNECTOR, 2nd anode: incls lead
	9D66133A06	CONNECTOR, cap: SS rectifier; less lead (pri/sec winding - USE 9C66133A11)
	15S10183A22	CONNECTOR, plug: 10 contact; less contacts
	15S10183A03	CONNECTOR, plug: 3 contact; less contacts (focus resistor)
	39S10184A02	CONTACT, plug: for 15S10183A22 connector
	39S10184A01	CONTACT, recept: for 15S10183A03 connector
	5K752248	GROMMET, tpg screw: plastic; H.V. transf mtg
	14C68711A06	INSULATOR CUP, for rect socket; incls wedge
	3S136942	SCREW, tpg: 6 x 3/8; rect socket mtg
	42A69711A01	SOCKET, solid state rectifier mtg

MISCELLANEOUS ELECTRICAL & MECHANICAL PARTS FOR C25 & AD25TS-915 (NOT ON PANELS)

ELECTRICAL PARTS
MISCELLANEOUS ELECTRICAL PARTS

REF.	PART NO.	DESCRIPTION
	1D69162A12	MOTOR & GEAR BOX, complete (AD25TS-915)
	1D69162A10	ARMATURE, motor: incls nylon gear (AD25TS-915)
	1D69162A14	GEAR BOX, only: less motor (AD25TS-915)
	1D69162A13	SWITCH, leaf: on gear box motor (AD25TS-915)
	40D69385A04	SWITCH, leaf: channel select; less housing, pushbutton, return spring & mtg nut (AD25TS-915)
	40D69385A03	SWITCH, leaf: intensity; incls pushbutton assembly (AD25TS-915)
	40D69385A03	SWITCH, leaf: hue; incls push-button assembly (AD25TS-915)

REF.	PART NO.	DESCRIPTION
	40D69385A03	SWITCH, leaf: volume; incls push-button assembly (AD25TS-915)
	40D66846A05	SWITCH, power on-off (AD25TS-915)
E-1C	80C68147A05	SPARK GAP
E-1R	48S134958	DIODE, silicon: D1H
E-1U	30V68641A87	MICROPHONE: incls cable & plug (AD25TS-915)
E-2C	80C68147A05	SPARK GAP
E-2G	48D67120A13	DIODE
E-2U	1V68623A17	SWITCH, slide: incls mtg brkt (remote/manual AD25TS-915)
E-3C	80C68147A05	SPARK GAP
E-4T	40V68617A74	SWITCH, AFT: incls mtg brkt (C25TS-915)
SW-4	40C70339A01	SWITCH, AFT (AD25TS-915)
E-5A	65S135685	LAMP, min incand: UHF/VHF; type #755
E-5B	65S135685	LAMP, min incand: UHF/VHF; type #755 (C25TS-915)
E-7J	51C67517A01	RES CAP
E-12J	48S137094	LINE TRANSIENT SUPPRESSOR: W1G
E-12X	65S10081A06	LAMP, glow: power on-off indicator type #3AD (AD25TS-915)
D-6A	48S137266	DIODE, Zener: D4H (AD25TS-915)

COILS & CHOKES

REF.	PART NO.	DESCRIPTION
L-1J	24C68976A02	CHOKE, AC line filter
L-5J	24V68639A80	COIL, CRT degaussing: complete; incls leads & connectors
L-9B	24D66772A12	CHOKE, 7.5 uh (RF AGC line - C25TS-915)
	24D68592B01	YOKE, deflection: 92° complete
	24D68717A04	CONVERGENCE YOKE, complete (C25TS-915)
	24D68717A08	CONVERGENCE YOKE, complete (AD25TS-915)
	24P65146A78	DYN CONV, incls cores: less radial magnet & spring (red)
	24P65146A78	DYN CONV, incls cores: less radial magnet & spring (grn)
	24P65146A78	DYN CONV, incls cores: less radial magnet & spring (blu)

TRANSISTORS

REF.	PART NO.	DESCRIPTION
Q-1H	48S134900	VERTICAL OUTPUT: A1C
Q-1R	48S134936	H.V. REGULATOR: A1Y
Q-5D	48S134920	AUDIO POWER AMP: A1N

CONTROLS

REF.	PART NO.	DESCRIPTION
R-1C	18D67502A12	G-2 grn: 10 meg
R-3C	18D67502A11	G-2 blue: 10 meg
R-6C	18D67502A10	G-2 red: 10 meg
R-7C	18D67502A13	FOCUS: 10 meg
R-8S	18D68443A19	INTENSITY: 1.5K (C25TS-915)
R-15C	18D67559A37	ABL: 1 meg
R-15D	18D68222A07	TONE: 50K
R-19E	18D68222A20	CONTRAST: 250Ω
R-24E	18D68222A06	VERTICAL HOLD: 100K
R-30D	18D68021A17	VOLUME: 20K; incls on-off switch (C25TS-915)
R-30L	18D68222A19	BRIGHTNESS: 750Ω
R-41A	18D67858A11	UHF RF DELAY: 100K (AD25TS-915)
R-59S	18D68443A18	HUE: 800Ω (C25TS-915)

TRANSFORMERS

REF.	PART NO.	DESCRIPTION
T-1A	24C68848A03	BALUN, UHF (located on antenna board - AD25TS-915)
T-1	24C68848A02	BALUN (located on antenna board - C25TS-915)
T-2J	25D68548A07	CRT FILAMENT (C25TS-915)
	25D68548A08	CRT FILAMENT (AD25TS-915)
T-3A	24C68848A02	BALUN, VHF (located on antenna board - AD25TS-915)
T-3D	25D68524A01	AUDIO OUTPUT

CTV5

REF.	PART NO.	DESCRIPTION	REF.	PART NO.	DESCRIPTION
MECHANICAL PARTS (NOT ON PANELS)					
	7D68564A09	BRACKET, CRT mtg: less rubber sleeve		29K580544	PLUG, 75Ω antenna input
				43C70407A04	SLEEVE, rubber: CRT mtg brkt
				41D65987A25	SPRING, CRT aquadag grnd

REF.	PART NO.	DESCRIPTION	REF.	PART NO.	DESCRIPTION
				3K749169	SCREW, 8-15 x 5/8 (CRT frame, slide rails & back cover spring mtg)
CABINET PARTS (WT676GW)				3S134013	SCREW, tpg: 8-15 x 1/2 (spkr grille retainer spring)
	13P65175A05	BEZEL, control panel: incls overlays & ctrl door; less speaker grille		3D66303A18	SCREW, tpg: 8-15 x 3/4 (CRT bezel to frame)
	13E70019A03	BEZEL, CRT		3D66303A19	SCREW, tpg: 8-15 x 13/16 (inter-lock)
	75B67953A02	BUMPER, cab foot		3S136869	SCREW, tpg: 8-18 X 1-1/8 phl flt blk ox (ctrl panel bezel to chassis)
	16E70084A07	CABINET, TV: metal; walnut (incls chassis slide rails)		50D67337A01	SPEAKER, 4 x 6 PM 16Ω VC
	42S10152A02	CLIP, push-on: spkr mtg		41A70069A02	SPRING, speaker grille retainer
	42S10122A01	CLIP, secures line cord to cab back			
	42C67787A04	CLIP, spring: cab back mtg		**IF- AUDIO PANEL "BA"**	
	30S183A13	CORD, AC line			
	15E70132D01	COVER, cabinet back: less all other components		1Y68662A57	IF-AUDIO PANEL "BA": complete
	15D68244A02	COVER, CRT: rear (USE 15D67156A02)	**ELECTRICAL PARTS**		
	15P65175A06	DOOR, control: incls overlay: less hinge spring clips			
	13E70068A01	GRILLE, speaker	**COILS & CHOKES**		
	42B70093A01	HINGE, spring clip: ctrl door (RH & LH)	L-1	24D68801A02	COIL, compensating: 450 uh
	36D69994A01	KNOB, channel selector: UHF (outer)	L-2	24D66772A12	CHOKE, resonant: 7.5 uh
	36D69994A02	KNOB, channel selector: VHF	L-3	24D68517A20	COIL, quadrature
	36D69982A02	KNOB, fine tune: UHF (inner)	L-4	24D68501A19	COIL, convertor secondary
	36D68218A35	KNOB, on-off vol & VHF fine tune	L-5	24D69707A13	COIL, hi pass filter: .326 uh
	36C69981A01	KNOB, slide: intensity & hue	L-6	24D68501A19	COIL, 39.75 MHz trap
	16B69964A01	LEG, cabinet	L-7	24D69707A13	COIL, hi pass filter: .326 uh
	13C69934A03	OVERLAY, control (on-vol VHF fine tune)	L-8	24D68501A16	COIL, 47.25 MHz trap
	13C69992A01	OVERLAY, control door (tone-bright-contrast-vertical)	L-9	24D68501A16	COIL, 41.25 MHz trap
	13C69993A02	OVERLAY, control panel (UHF/VHF intensity & hue)	L-11	24D66772A12	CHOKE, resonant: 7.5 uh
	13C69934A01	OVERLAY, nameplate (Motorola)	L-12	24D68588A05	COIL, 4.5 MHz trap & CTO
	64B69949A01	PLATE, metal: cab leg	L-13	24D68517A19	COIL, 4.5 MHz audio take-off
	34D69996A02	SCALE, dial: UHF			
	34D6 9996A01	SCALE, dial: VHF	**INTEGRATED CIRCUITS**		
	3S136802	SCREW, tpg: 8-18 x 3/8 phl trs (cab leg plate mtg)	IC-1	51D70177A02	INTEGRATED CIRCUIT: IF sound
	3S134013	SCREW, tpg: 8 x 1/2 (spkr grille retainer spring mtg)	**DIODES**		
	3D66303A18	SCREW, tpg: 8-15 x 3/4; (CRT bezel to cab)	D-1	48D67120A11	DIODE, low power
	3D66303A19	SCREW, tpg: 8-15 x 13/16 (interlock)	D-2	48C65837A02	DIODE, crystal
	3S136869	SCREW, tpg: 8-18 x 1-1/8 phl flt blk ox (ctrl panel bezel to chassis)	D-3	48C65837A02	DIODE, crystal
	50D68384A02	SPEAKER, 4" PM 16Ω VC	D-4	48S137133	DIODE, silicon: D3A
	41A70069A01	SPRING, speaker grille retainer	**TRANSISTORS**		
			Q-1	48S137127	RF AGC DELAY: P2S
			Q-2	48S134933	AGC AMP: A1V
CABINET PARTS (WU819GW)			Q-3	48S134815	AGC KEYER: M4815 (USE 48-134910)
	13P65175A07	BEZEL, control panel: incls overlays & ctrl door; less speaker grille	Q-4	48S137127	AUDIO DRIVER: P2S
	13E69927A04	BEZEL, CRT	Q-5	48S137169	AUDIO OUTPUT: A6G
	16E70255A01	CABINET, TV: console; walnut	Q-6	48S134981	1ST IF INTERSTAGE: A2Y (USE 48-134904)
	42S10152A02	CLIP, push-on: spkr mtg	Q-7	48S134932	2ND IF INTERSTAGE: A1U
	42S10122A01	CLIP, secures line cord to ant cover on cab back	Q-8	48S134932	3RD IF: A1U
	42C67787A04	CLIP, spring: cab back mtg	Q-9	48S137172	1ST VIDEO: A6J
	15D70134B01	COVER, ant terminal board: located on back cover (polystyrene)	Q-10	48S137107	AUDIO AMP: A5M
	15P65173A57	COVER, cabinet back: incls ant terminal cover; less all other components	Q-11	48S137168	AUDIO OUTPUT: P2V
	15E68244A03	COVER, CRT: rear	**TRANSFORMERS**		
	30S183A13	CORD, AC line	T-1	24D68501A17	1ST IF INTERSTAGE
	15P65175A08	DOOR, control: incls overlay; less hinge spring clip	T-2	24D68501A20	2ND IF INTERSTAGE
	55C63659A28	GLIDE, dome: cab leg	T-3	24D68588A04	3RD IF
	13E69943A03	GRILLE, speaker: less mtg spring	**CONTROLS**		
	42B70093A01	HINGE, spring clip: control door (R & L)	R-14	18D66401A36	AGC SET-UP: 10K
	36D69994A01	KNOB, channel selector: UHF (outer)	R-27	18D66401A22	47.25MHz TRAP ADJ: 100 ohms
	36D69994A02	KNOB, channel selector: VHF	R-38	18D66401A22	41.25MHz TRAP ADJ: 100 ohms
	36D69982A02	KNOB, fine tune: UHF (inner)	**MECHANICAL PARTS**		
	36D68218A35	KNOB, on-off vol & VHF fine tune		29S10134A29	CONNECTOR, recept: panel mtg
	36C69981A01	KNOB, slide: intensity & hue		29S10134A32	LUG, terminal: AFC B+ take-off
	16B70115A01	LEG, cabinet		9C67349B04	RECEPTACLE, IF input
	13C69934A03	OVERLAY, control (on-vol VHF fine tune)			
	13C69992A01	OVERLAY, control door (tone-bright-contrast-vertical)		**DEFLECTION PANEL "FA"**	
	13D69993A02	OVERLAY, control panel (UHF/VHF intensity & hue)		1Y68662A56	DEFLECTION PANEL "FA": complete
	13C69934A01	OVERLAY, nameplate (Motorola)	**ELECTRICAL PARTS**		
	64A67158A02	PLATE, metal: leg mtg	**DIODES**		
	34D69996A02	SCALE, dial: UHF			
	34D69996A01	SCALE, dial: VHF	D-1	48S137167	PHASE DETECTOR, dual; D3G

REF.	PART NO.	DESCRIPTION
COILS & CHOKES		
L-1	24D68778A01	HORIZ OSCILLATOR, incls core; less adjustment rod
TRANSISTORS		
Q-1	48S137172	BURST GATE FORMER: A6J
Q-2	48S137173	SYNC SEPARATOR: P2W
Q-3	48S137172	HORIZ OSCILLATOR: A6J
CONTROLS		
R-26	18D67678A11	VERTICAL LINEARITY, 3 meg incls R31 vert size
R-31	18D67678A11	VERTICAL SIZE, 100K incls R26 vert lin.
RESISTORS		
R-54	6C66263A18	VARISTOR
MECHANICAL PARTS		
	29S10134A29	CONNECTOR, recept: panel mtg
	47C66082A04	ROD, adjustment: horiz osc (L1)
	26D70088A06	SHIELD, horiz osc coil: L1
	9C67618A05	SOCKET, tube: V3
	9C67582A02	SOCKET, tube: V4

CONVERGENCE PANEL "HA"

REF.	PART NO.	DESCRIPTION
	1Y68662A55	CONVERGENCE PANEL "HA": complete

ELECTRICAL PARTS

REF.	PART NO.	DESCRIPTION
DIODES		
D-1	48S191A08	RECTIFIER, silicon
D-2	48S191A08	RECTIFIER, silicon
D-3	48S10062A01	RECTIFIER, silicon (USE 48-191A08)
COILS & CHOKES		
L-1	24V68609A47	COIL, R/G right side vert lines: incls mtg nut
L-2	24D67682A06	COIL, R/G horiz lines
L-3	24D67682A03	COIL, blue horiz tilt (USE 24-67682A11)
L-4	24D67682A03	COIL, blue cent horiz phase (USE 24-67682A11)
CONTROLS		
R-1	18D67671A07	BLUE VERT TILT: 200Ω
R-2	18D67671A07	R/G VERT TILT: 200Ω
R-3	18D67671A05	R/G VERT DIFF AMP: 500Ω
R-4	18D67671A01	R/G VERT DIFF TILT: 30Ω
R-6	18D67671A01	BLUE VERT AMP: 30Ω
R-7	18D67671A09	R/G VERT AMP, 120Ω (some sets used 200Ω control) replace with 120Ω control and remove 270Ω resistor (R15)
R-9	18D67671A14	R/G - L.S. VERT LINES: 90Ω
R-12	18D67671A11	R/G HORIZ DIFF TILT: 150Ω
R-13	18D67671A04	BLUE HORIZ AMP: 150Ω
MECHANICAL PARTS		
	29S10134A29	CONNECTOR, recept: panel mtg
	29S10134A33	LUG, connector: jumper lead
	29S10134A32	LUG, terminal: jumper lead
	2C720979	NUT, coil mtg: (L1)
	9C67580A03	SOCKET, 12 pin

AFT PANEL "KA"

REF.	PART NO.	DESCRIPTION
	1Y68662A60	AFT PANEL "KA": complete

ELECTRICAL PARTS

REF.	PART NO.	DESCRIPTION
COILS & CHOKE		
L-1	24D66772A12	CHOKE, RF
L-2	24D66772A12	CHOKE, RF

REF.	PART NO.	DESCRIPTION
TRANSFORMERS		
T-1	24D68501A21	AFT DISCRIMINATOR
DIODES		
D-1	48C65837A02	CRYSTAL
D-2	48C65837A02	CRYSTAL
TRANSISTORS		
Q-1	48S134937	EMITTER FOLLOWER: A1Z
Q-2	48S134932	DISCRIMINATOR: A1U
MECHANICAL PARTS		
	42S10152A07	CLIP, push-on: AFT panel mtg
	15S10183A15	CONNECTOR, plug: 1 contact; less contact AFT output
	39S10184A02	CONTACT, plug: for 1 pin connector plug (AFT output)
	29S10134A15	LUG, connector: AFT input
	29S10134A33	LUG, connector: B+
	15S70096A01	SHIELD, AFT panel: less panel mtg clips

COLOR - VIDEO PANEL "SA"

REF.	PART NO.	DESCRIPTION
	1Y68662A59	COLOR VIDEO PANEL "SA": complete

ELECTRICAL PARTS

REF.	PART NO.	DESCRIPTION
INTEGRATED CIRCUIT "IC"		
IC	51D70177A01	COLOR DEMODULATOR
CRYSTAL		
Y-1	48C66865A04	3.58 MHz crystal
COILS & CHOKES		
L-1	24D69708A02	CHOKE, 1st color IF
L-2	24D68852A03	DELAY LINE
L-3	24D68801A30	COMPENSATING: 25 uh
L-4	24D68801A46	COMPENSATING: 800 uh
L-5	24D68517A12	3.58 MHz oscillator
L-6	24D68002A98	COMPENSATING: 5.6 uh
L-7	24D68517A21	3.58 MHz limiter
L-8	24D68801A30	COMPENSATING: 25 uh
L-9	24D68801A48	COMPENSATING: 8.2 uh
L-10	24D68801A48	COMPENSATING: 8.2 uh
L-11	24D68801A48	COMPENSATING: 8.2 uh
L-12	24D68801A03	COMPENSATING: 100 uh
TRANSFORMERS		
T-1	24D68517A18	2ND COLOR IF
DIODES & TRANSISTORS		
D-1	48G10346A02	DIODE (USE 48-67120A11)
D-3	48S191A04	RECTIFIER, silicon (USE 48-191A07)
D-4	48S137133	DIODE: D3A
D-5	48G10346A01	DIODE (USE 48-67120A11)
D-7	48G10346A01	DIODE (USE 48-67120A11)
D-8	48S137133	DIODE: D3A
Q-1	48S134841	1ST COLOR IF: M4841
Q-2	48S134841	2ND COLOR IF: M4841
Q-3	48S134841	2ND VIDEO AMP: M4841
Q-4	48S137113	VOLTAGE REGULATOR: A5T
	48S137002	VOLTAGE REGULATOR: A3M (in panel SA-3 & later)
Q-5	48S134842	PULSE LIMITER & INVERTER: M4842
Q-6	48S137111	COLOR SYNC GATE & AMP: A5S
Q-7	48S134842	CRYSTAL DRIVER: M4842
Q-8	48S134841	CRYSTAL AMP: M4841
Q-9	48S137115	ACC AMP: A5U
Q-10	48S137003	CRYSTAL OSCILLATOR: A3N
Q-11	48S137127	COLOR KILLER: P2S
Q-12	48S134970	PHASE SPLITTER: A2T
Q-13	48S134841	PHASE SHIFTER: M4841
Q-14	48S134918	3.58 LIMITER: A1L
Q-15	48S137002	RED VIDEO OUTPUT: A3M
Q-16	48S137002	BLUE VIDEO OUTPUT: A3M
Q-17	48S137002	GREEN VIDEO OUTPUT: A3M
Q-18	48S137127	SYNC & AGC TAKE-OFF: P2S
CONTROLS		
R-76	18D66401A42	BLUE DRIVE: 100Ω

REF.	PART NO.	DESCRIPTION
R-83	18D66401A41	GREEN DRIVE: 100Ω
R-89	18D66401A40	RED DRIVE: 100Ω

MECHANICAL PARTS

REF.	PART NO.	DESCRIPTION
	29S10134A29	CONNECTOR, recept: panel mtg
	26C66745A07	HEAT SINK, for transistor Q4 (A5T type)
	26C66745A05	HEAT SINK, for transistors Q15, 16 & 17 (also Q4 A3M type)
	29S10134A37	LUG, terminal; video output
	43B68719A01	SPACER, for transistors Q15, 16 & 17 (also Q4 A3M type)

DC REGULATOR PANEL ZA

REF.	PART NO.	DESCRIPTION
	1Y68662A58	DC REGULATOR PANEL "ZA": complete

ELECTRICAL PARTS

DIODES

REF.	PART NO.	DESCRIPTION
D-1	48S191A05	RECTIFIER, silicon (USE 48-191A07)

TRANSISTORS

REF.	PART NO.	DESCRIPTION
Q-1	48S137172	OVERLOAD PROTECTOR: A6J
Q-2	48S134842	REFERENCE AMP: M4842 (USE 48-134992)
Q-3	48S137170	REFERENCE DIODE, silicon Zener; D3H

CONTROLS

REF.	PART NO.	DESCRIPTION
R-2	18D66401A33	VOLT REGULATOR ADJUST: 2K

MECHANICAL PARTS

REF.	PART NO.	DESCRIPTION
	29S10134A29	CONNECTOR, recept: panel mtg

CHASSIS ELECTRICAL PARTS (NOT ON PANELS)

MISCELLANEOUS ELECTRICAL PARTS

REF.	PART NO.	DESCRIPTION
CB-800	80C66390A20	CIRCUIT BREAKER
PL-800	65S135685	LAMP, min incand: 6.3V - .15A #755 (UHF)
PL-801	65S135685	LAMP, min incand: 6.3V - .15A #755 (VHF)
RC-800	51C67517A01	RES CAP
SW-7	40D69977A01	SWITCH, slide: AFT
SW-800	- - -	SWITCH, on-off (part of 301 vol control)

DIODES

REF.	PART NO.	DESCRIPTION
D-200	48D67120A02	DIODE
D-500	48D69723A01	RECTIFIER, solid state HV
D-600	48D67120A11	DIODE
D-800	48S191A05	RECTIFIER, silicon (USE 48S191A07)
D-801	48S191A05	RECTIFIER, silicon (USE 48S191A07)

TRANSISTORS

REF.	PART NO.	DESCRIPTION
Q-800	48S134936	REGULATOR: A1Y

COILS & CHOKES

REF.	PART NO.	DESCRIPTION
L-100	24D68002A66	COMPENSATING: 220 uh
L-200	24D66772A03	CHOKE: 6.6 uh
L-201	24D66772A03	CHOKE: 6.6 uh
L-202	24D66772A03	CHOKE: 6.6 uh
L-500	24C70169A01	COIL, phasing: PCC
L-501	24D66772A03	CHOKE: 6.6 uh
L-502	24D66772A03	CHOKE: 6.6 uh
L-503	- - -	PART OF PLATE CAP ASSEMBLY (see mechanical parts list)
L-701	24G10250A11	DEFLECTION YOKE: 92° (19" CRT)
	24G10250A10	DEFLECTION YOKE: 92° (21" & 25" CRT)
L-702	24P65146A78	DYNAMIC CONVERGENCE COIL, incls core; less radial magnet & spring
L-703	24P65146A78	DYNAMIC CONVERGENCE COIL, incls core; less radial magnet & spring
L-704	24P65146A78	DYNAMIC CONVERGENCE COIL, incls core; less radial magnet & spring
	24D68717A07	CONVERGENCE YOKE, complete
L-800	25D67554A17	CHOKE, filter

REF.	PART NO.	DESCRIPTION
L-801	24V68636A27	COIL, CRT degaussing: complete (19" & 21" CRT)
	24V68636A98	COIL, CRT degaussing: complete (25" CRT)
L-804	24C68976A02	CHOKE, line filter

TRANSFORMERS

REF.	PART NO.	DESCRIPTION
T-500	25D70067A01	VERT PINCUSHION SATURABLE REACTOR
T-501	24D67564A08	HORIZONTAL OUTPUT TRANSFORMER, complete; less solid state rectifier & socket assembly
T-600	25V68636A65	VERTICAL OUTPUT, incls 3 pin connector with contact
T-800	25V68636A39	POWER, incls 3 & 12 pin connectors with contacts

CONTROLS

REF.	PART NO.	DESCRIPTION
R-100	18D67858A07	AGC DELAY, 1KΩ
R-104	18D68221A08	CONTRAST & BLUE G2, contrast 750Ω blue G2 3 mge
R-108	18D67858A10	VIDEO PEAKING, 3.5K
R-206	18D68221A09	BRIGHTNESS & RED G2, brightness 1KΩ, red G2 3 meg
R-215	18D67502A20	FOCUS, 10 meg
R-300	18D68221A07	MASTER BRIGHTNESS & TONE, master brightness 3.5KΩ, tone 5KΩ
R-301	18D68021A15	VOLUME & ON-OFF, vol 50K; incls (SW800) on-off switch
R-503	18P65173A87	HORIZ BIAS, 180KΩ; SPECIAL
R-603	18D68221A10	VERTICAL HOLD & GRN G2, vert hold 1 meg, grn G2 3 meg
R-900	18D68443A17	INTENSITY, 5KΩ
R-901	18D68443A16	HUE, 25KΩ

CHASSIS MECHANICAL PARTS (NOT ON PANELS)

REF.	PART NO.	DESCRIPTION
	14D68726A07	ANTENNA BOARD, incls ant terminals & 72Ω socket; less balum coil
	7D70099A01	BRACKET, CRT mtg: less cloth tape; 21" CRT
	7D68564A09	BRACKET, CRT mtg: less cloth tape; 25" CRT
	7D67151A05	BRACKET, conv yoke: plastic; coil mtg
	43C67077A05	BUSHING, snap-in: 2nd anode lead dress
	43C67077A03	BUSHING, snap-in: lead dress; H. V. transf (ID 5/8")
	30V68636A89	CABLE, shielded: incls plug (ant to VHF tuner)
	30D67870A14	CABLE, shielded: incls plugs (UHF to VHF tuner)
	30D70172A02	CABLE, shielded: incls plugs (convertor)
	9V68636A38	CAP, plate: incls wire & choke (L503) V1 horiz output
	42C65572A15	CLAMP, degausser coil mtg: 19" CRT
	42C69663A01	CLAMP, degausser coil mtg (1-3/4" long - loop on one end) 21" & 25" CRT
	42C69663A02	CLAMP, degausser coil mtg (3" long - loop on both ends) 21" & 25" CRT
	42C69663A03	CLAMP, degausser coil mtg (3-1/8" long - loop on one end) 21" & 25" CRT
	42A70203A01	CLAMP, metal: IF/audio panel mtg
	42S10240A05	CLIP, metal: lytic mtg (C800)
	42S10240A04	CLIP, metal: lytic mtg (C801 & C803)
	42A70106A01	CLIP, metal: module panel grnd
	42A69711A01	CLIP, metal: solid state rect; (located inside insulator cup)
	42B67815B03	CLIP, wire: nylon; lead dress
	42D65864A46	CONNECTOR, 2nd anode: incls lead
	31C68421A03	CONNECTOR, module panel: 5 pin; on chassis
	31D70080A02	CONNECTOR, module panel: 6 pin; on chassis
	15S10183A15	CONNECTOR, plug: 1 contact; less contact (AFT)
P-1	15S10183A10	CONNECTOR, plug: 8 contact; less contact (def yoke)
P-2	15S10183A20	CONNECTOR, plug: 12 contact; less contacts (power transf)
P-3	15S10183A14	CONNECTOR, plug: 10 contact; less contacts (pulse)

REF.	PART NO.	DESCRIPTION
P-4	15S10183A04	CONNECTOR, plug: 3 contact; less contacts (CRT filament)
P-5	15S10183A15	CONNECTOR, plug: 1 contact; less contacts (focus resistor)
P-6	15S10183A12	CONNECTOR, plug: 6 contact; less contacts (vert output transf & G2 control)
P-7	15S10183A12	CONNECTOR, plug: 6 contact; less contacts (VHF tuner)
P-8	15S10183A04	CONNECTOR, plug: 3 contact; less contacts (UHF to VHF tuner)
	15S10183A16	CONNECTOR, recept: 1 contact; less contact (AFT)
S-1	15S10183A09	CONNECTOR, recept: 8 contact; less contacts (def yoke)
S-2	15S10183A19	CONNECTOR, recept: 12 contact; less contacts (power transf)
S-3	15S10183A13	CONNECTOR, recept: 10 contact; less contacts (pulse)
S-4	15S10183A03	CONNECTOR, recept: 3 contact; less contacts (CRT filament)
S-5	15S10183A16	CONNECTOR, recept: 1 contact; less contact (focus resistor)
S-6	15S10183A11	CONNECTOR, recept: 6 contact; less contacts (vert output transf & G2 control)
S-7	15S10183A11	CONNECTOR, recept: 6 contact; less contacts (VHF tuner)
S-8	15S10183A03	CONNECTOR, recept: 3 contact; less contacts (UHF to VHF tuner)
	39S10184A02	CONTACT, plug (for focus & P1 thru P8 connectors)
	39S10184A01	CONTACT, recept (for focus & S1 thru S8 connectors)
	44D67370A10	GEAR, idler: VHF fine tune
	44D67370A12	GEAR & SHAFT: VHF fine tune
	5D69911A01	GROMMET, plastic: conv panel module mtg (only)
	5D69911A02	GROMMET, plastic: module panel mtg
	5C736116	GROMMET, screw: plastic; interlock
	14C68711A06	INSULATOR, cup: solid state rect; incls wedge
	14A63948A01	INSULATOR, mica; transistor socket (Q800 regulator)
	14B65732A04	INSULATOR, terminal lug: (for 9C66402A04)
	36A69907A01	KNOB, radial magnet: conv yoke
	29S10134A27	LUG, terminal: UHF ant leads
	9C66402A05	LUG, terminal: spkr leads
	29S10134A33	LUG, terminal: video drive terminal
	9C66402A04	LUG, terminal: master G1 taps
	59G10296A01	MAGNET, blue lateral & purity ring
	76D66816A02	MAGNET, radial: conv yoke; less knob
	2S7051	NUT, hex: 3/8-32; ctrl mtg
	2S10054A56	NUT, spring: clip-on; yoke ring
	2S132377	NUT, wing: 6-32; conv yoke
	28B66635A08	PLUG, AC interlock
	38B70116A01	PLUG, AFT switch; blk
	28C67679A04	PLUG, molded: 12 pin; conv yoke
	42C67369A02	RETAINER, gear: VHF fine tune gears; plastic
	42D70108A01	RING, metal: yoke retainer; 19" CRT
	42D70108A03	RING, metal: yoke retainer; 21" & 25" CRT
	5S6846	RIVET (chassis slide lock spring mtg)
	5S10281A02	RIVET, drive pin: plastic; (hinge for ctrl brkt-USE 28-10114A10)
	3B69287A03	SCREW, thumb: def yoke
	3S132346	SCREW, mch: 6-32 x 2-1/8; conv yoke
	3B68012A03	SCREW, mch: 1/4-20 x 5/8; CRT mtg (19" CRT)
	3B68012A02	SCREW, mch: 1/4-20 x 7/8; CRT mtg (21" & 25" CRT)
	3B70184A01	SCREW, tpg: 6-20 x 5/16 ; VHF tuner mtg
	3S136942	SCREW, tpg: 6-32 x 3/8; solid state rect clip mtg
	3S121599	SCREW, tpg: 8-32 x 3/4 (gear retainer)
	26C65584B05	SHIELD, metal: UHF/VHF pilot light
	9V68636A70	SOCKET, CRT: complete; incls leads lugs, plugs & all electrical components in leads

REF.	PART NO.	DESCRIPTION
	9V68636A83	SOCKET, pilot light: UHF & VHF; incls lead & contact
	9D66770A19	SOCKET, tube: 9 pin; V2 damper
	9D66770A17	SOCKET, tube: 12 pin; V1 horiz output
	9C67532A02	SOCKET, transistor: Q800 regulator
	43C70189A01	SPACER, metal: yoke ring insert (21" & 25" CRT) some sets
	41A65351A05	SPRING, back cover ground
	41B70094A01	SPRING, chassis slide: lock
	41B70164B01	SPRING, chassis slide: stop
	41C65987A24	SPRING, CRT aquadag grnd (19" CRT)
	41C65987A23	SPRING, CRT aquadag grnd (21" & 25" CRT)
	41B69152A01	SPRING, retainer: CRT shield mtg (19" & 25" CRT)
	41C65597A02	SPRING, retainer: radial magnet; conv yoke
	42C65572A05	STRAP, adjustable: degausser coil; 21" CRT
	42D67027A04	STRAP, metal: CRT mtg; 21" CRT
	42D67027A03	STRAP, metal: CRT mtg; 25" CRT

VHF TUNER OPTT-429

ELECTRICAL PARTS

MISCELLANEOUS ELECTRICAL PARTS

REF.	PART NO.	DESCRIPTION
D-1	48P65173A77	DIODE, AFT: BA142-01
E-1	24P65173A80	ANTENNA INPUT FILTER: complete
SW-6	40P65173A79	SWITCH, UHF B+ and pilot light

COILS & CHOKES

L-22	24P65148A05	COIL, master oscillator: incls core
L-23	24P65173A82	COIL, mixer: incls core

TRANSISTORS

Q-1	48P65146A61	RF AMP: SE5020
Q-2	48P65173A78	MIXER: SE5030
Q-3	48P65146A63	OSCILLATOR: SE1010
T-1	24C68848A02	BALUN, transformer (located on antenna terminal board)

MECHANICAL PARTS

REF.	PART NO.	DESCRIPTION
	77V68636A78	VHF TUNER, complete; incls plugs, convertor, cable, mtg brkt and all other outboard component parts less ant input & VHF to UHF shielded cables.
1	1P65120A46	DRIVE LINK ASSEMBLY, incls 10 & 14 tooth gears
2	47P65146A44	GEAR, fine tune: incls bushing, clutch & metal gear
3	44P65146A15	GEAR, fine tune: 14 tooth; engages gear head screws
4	44P65146A16	GEAR, idler: 10 tooth; drive link
5	3P65145A82	SCREW, gear head: fine tune
6	47P65173A81	SHAFT, channel selector: incls detent wheel & pre-set screw & coil assembly
7	42P65120A43	SPRING, bias: retains shaft to front plate
8	41A747735	SPRING, chan sel shaft grounding
9	42P65145A84	SPRING, detent: chan sel shaft (located on front plate)
10	41P65120A42	SPRING, drive link return
11	4P65114A53	WASHER, "C" (secures fine tune gear to chan sel shaft)

VHF TUNER OPTT -430

ELECTRICAL PARTS

MISCELLANEOUS ELECTRICAL PARTS

REF.	PART NO.	DESCRIPTION
D-1	48X90233A08	DIODE, AFT: S2085
E-1	24P65171A94	ANTENNA INPUT ASSEMBLY
E-2	76X90301A01	FERRITE BEAD
SW-1C	59X90285A03	OSCILLATOR STATOR ASSEMBLY
SW-6	40P65149A66	SWITCH, UHF B+

COILS & CHOKES

L-14	24X90243A32	COIL, UHF INPUT, shunt
L-15	24X90243A33	COIL, UHF INPUT, series

REF.	PART NO.	DESCRIPTION
L-24	24P65149A41	COIL, mixer: incls core
L-40	24X90243A34	COIL, FM trap: incls core & C6
L-45	24X90243A31	CHOKE, RF

TRANSISTORS

REF.	PART NO.	DESCRIPTION
Q-1	48S137158	RF: A6E
Q-2	48S134950	MIXER: A2H
Q-3	48S134949	OSCILLATOR: A2G

TRANSFORMER

REF.	PART NO.	DESCRIPTION
T-1	24C68848A02	BALUN (located on antenna terminal board)

MECHANICAL PARTS

REF.	PART NO.	DESCRIPTION
	77V68636A78	VHF TUNER, complete; incls plugs, convertor cable, mtg brkt and all other outboard component parts, less antenna input & UHF to VHF shielded cables
1	47X90276A05	FRONT PLATE ASSEMBLY, complete: incls chan sel shaft fine tune, idler & drive gears & pre-set screw & coil assembly (SW1A &B)
2	44X90277A02	GEAR, fine tune: incls cam assembly
3	3X90262A04	SCREW, gear head: fine tune
4	41X90271A07	SPRING, channel selector shaft ground
5	4X90259A05	WASHER, "C": secures fine tune gear to chan sel shaft

UHF TUNER TT-645

ELECTRICAL PARTS

REF.	PART NO.	DESCRIPTION
D-1	48P65112A73	DIODE, mixer
D-2	48P65148A02	DIODE, AFC
Q-1	48P65174A24	TRANSISTOR: SPS4145

MECHANICAL PARTS

REF.	PART NO.	DESCRIPTION
	77P65174A11	TUNER, TT645; complete

UHF TUNER TT-649

ELECTRICAL PARTS

REF.	PART NO.	DESCRIPTION
D-1	48P65112A73	DIODE, mixer
D-2	48X90233A09	DIODE, AFT: 1S1923A
Q-1	48S134902	TRANSISTOR: A1E

MECHANICAL PARTS

REF.	PART NO.	DESCRIPTION
	77P65174A11	TUNER, TT649; complete

CABINET PARTS MODEL WT561FW

REF.	PART NO.	DESCRIPTION
	85D67625A06	ANTENNA, UHF: "bow tie"
	85D68375A02	ANTENNA, VHF: dipole; complete
	13P65173A49	BEZEL, control panel: incls ctrl door, ctrl esc & overlays; less speaker grille
	13E70020B02	BEZEL, CRT
	75B67953A02	BUMPER, cabinet foot
	16E70084A03	CABINET, TV: metal; walnut (incls chassis slide rails)
	42S10122A01	CLIP, secures line cord to cab back
	42S10152A02	CLIP, push-on: spkr mtg
	42C67787A04	CLIP, spring: cab back mtg
	30S183A13	CORD, AC line
	15E70132A01	COVER, cabinet back: less all other components
	15D68244A04	COVER, CRT: rear
	15P65173A52	DOOR, control: incls ctrl overlay; less hinge spring clips
	13P65173A50	ESCUTCHEON, control panel: UHF/VHF; incls slide ctrl overlay
	13E70068A01	GRILLE, speaker: less mtg spring
	42B70093A01	HINGE, spring clip: ctrl door (RH & LH)
	36D69994A01	KNOB, channel selector: UHF (outer)

REF.	PART NO.	DESCRIPTION
	36D69994A02	KNOB, channel selector: VHF
	36D69982A02	KNOB, fine tune: UHF (inner)
	36D68218A35	KNOB, on-off vol & VHF fine tune
	36C69981A01	KNOB, slide: intensity & hue
	13C69934A04	OVERLAY, control(on-vol VHF fine tune)
	13C69992A02	OVERLAY, control door (tone-bright-contrast-vertical)
	13C69934A02	OVERLAY, nameplate (Motorola)
	13C69945A03	OVERLAY, slide control (intensity hue)
	34D69996A03	SCALE, dial: UHF
	34D69996A04	SCALE, dial: VHF
	3S134013	SCREW, tpg: 8 x 1/2; spkr grille retainer spring mtg
	3S134014	SCREW, tpg: 8 x 5/8; VHF ant mtg
	3S136869	SCREW, tpg: 8-18 x 1-1/8 (phl flt blk ox) ctrl panel bezel to chassis
	3D66303A18	SCREW, tpg: 8-15 x 3/4; CRT bezel to cab
	3D66303A19	SCREW tpg: 8-15 x 13/16; interlock
	50D68384A02	SPEAKER, 4" PM 16Ω VC
	41B65987B09	SPRING, CRT bezel ground
	41A70069A01	SPRING, speaker grille retainer

CABINET PARTS MODEL WT678FW

REF.	PART NO.	DESCRIPTION
	85D67625A06	ANTENNA, UHF: "bow tie"
	85D68375A02	ANTENNA, VHF: dipole; complete
	13P65173A56	BEZEL, control panel: incls ctrl door, ctrl esc & overlays; less speaker grille
	13E70019A02	BEZEL, CRT
	75B67953A02	BUMPER, cabinet foot
	16E70084A07	CABINET, TV: metal; walnut (incls chassis slide rails)
	42S10122A01	CLIP, secures line cord to cab back
	42S10152A02	CLIP, push-on: spkr mtg
	42C67787A04	CLIP, spring: cab back mtg
	30S183A13	CORD, AC line
	15E70132A01	COVER, cabinet back: less all other components
	15D68244A02	COVER, CRT: rear (USE 15-67156A02)
	15P65173A54	DOOR, control: incls cloth panel & ctrl overlays; less hinge spring clips
	13P65173A55	ESCUTCHEON, control panel: UHF/VHF/ incls slide control overlay
	13P65173A53	GRILLE, speaker: incls cloth panel overlay; less mtg spring
	42B70093A01	HINGE, spring clip: ctrl door (RH & LH)
	36D69994A01	KNOB, channel selector: UHF (outer)
	36D69994A02	KNOB, channel selector: VHF
	36D69982A02	KNOB, fine tune: UHF (inner)
	36D68218A35	KNOB, on-off vol & VHF fine tune
	36C69981A01	KNOB, slide: intensity & hue
	13C69934A03	OVERLAY, control: (on-vol VHF fine tune)
	13C69992A03	OVERLAY, control door (tone-bright-contrast-vertical)
	13C69957A03	OVERLAY, control door (cloth panel)
	13C69934A01	OVERLAY, nameplate (Motorola)
	13C69945A01	OVERLAY, slide control (intensity hue)
	13C69957A04	OVERLAY, speaker grille (cloth panel)
	34D69996A03	SCALE, dial: UHF
	34D69996A04	SCALE, dial: VHF
	3S134013	SCREW, tpg: 8 x 1/2 (spkr grille retainer spring mtg)
	3S134014	SCREW, tpg: 8 x 5/8; VHF ant mtg
	3S136869	SCREW, tpg: 8-18 x 1-1/8 (phl flt blk ox) ctrl panel bezel to chassis
	3D66303A18	SCREW, tpg: 8-15 x 3/4; CRT bezel to cab
	3D66303A19	SCREW, tpg: 8-15 x 13/16; interlock
	50D68384A02	SPEAKER, 4" PM 16Ω VC
	41B65987B09	SPRING, CRT bezel ground
	41A70069A01	SPRING, speaker grille retainer

CABINET PARTS MODEL WT815GW

REF.	PART NO.	DESCRIPTION
	85D67625A06	ANTENNA, UHF; "bow-tie"
	85D68375A02	ANTENNA, VHF; dipole: complete
	13P65173A60	BEZEL, control panel: incls ctrl door, ctrl esc & overlays; less ctrl door cloth panel overlay & spkr grille
	13E69927A01	BEZEL, CRT
	75B67953A02	BUMPER, cabinet foot
	16E70107A01	CABINET, TV; metal: walnut (incls chassis slide rails)
	42S10122A01	CLIP, secures line cord to ant cover on cab back
	42S10152A02	CLIP, push-on: spkr mtg
	42C67787A04	CLIP, spring: cab back mtg
	15D70134A01	COVER, ant terminal board: located on back cover (polystyrene)
	15P65174A13	COVER, cab back: incls ant terminal cover; less all other components
	15D68244A03	COVER, CRT: rear
	30S183A13	CORD, AC line
	15P65173A59	DOOR, control: incls ctrl overlay; less cloth panel overlay & hinge spring clips
	13P65173A55	ESCUTCHEON, control panel: UHF/ VHF/ incls slide ctrl overlay
	13E69943A02	GRILLE, speaker: less mtg spring & cloth panel overlay
	42B70093A01	HINGE, spring clip: control door (R & L)
	36D699 94A01	KNOB, channel selector: UHF (outer)
	36D699 94A02	KNOB, channel selector: VHF
	36D69982A02	KNOB, fine tune: UHF (inner)
	36D68218A35	KNOB, on-off vol & VHF fine tune
	36C69981A01	KNOB, slide: intensity & hue
	13C69934A03	OVERLAY, control (on-vol VHF fine tune)
	13C69992A03	OVERLAY, control door (tone-bright-contrast-vertical)
	13D69957A01	OVERLAY, control door (cloth panel)
	13C69934A01	OVERLAY, nameplate (Motorola)
	13C69945A01	OVERLAY, slide control: intensity hue
	13D69957A02	OVERLAY, speaker grille (cloth panel)
	34D69996A04	SCALE, dial: UHF
	34D69996A03	SCALE, dial: VHF
	3S134014	SCREW, tpg: 8 x 5/8; VHF ant mtg
	3S134013	SCREW, tpg: 8-15 x 1/2 (spkr grille retainer spring)
	3D66303A18	SCREW, tpg: 8-15 x 3/4 (CRT bezel to frame)
	3D66303A19	SCREW, tpg: 8-15 x 13/16 (interlock)
	3S136869	SCREW, tpg: 8-18 x 1-1/8 (phl flt blk ox) ctrl panel bezel to chassis
	50D67337A01	SPEAKER, 4 x 6 PM 16Ω VC
	41A70069A01	SPRING, speaker grille retainer

CABINET PARTS MODEL WU817GW

REF.	PART NO.	DESCRIPTION
	13P65173A60	BEZEL, control panel: incls ctrl door, ctrl esc & overlays; less ctrl door cloth panel overlay & spkr grille
	13E69927A02	BEZEL, CRT
	16E70255A01	CABINET, TV: console; walnut
	42S10122A01	CLIP, secures line cord to ant cover on cab back
	42S10152A02	CLIP, push-on: spkr mtg
	42C67787A04	CLIP, spring: cab back mtg
	15D70134A01	COVER, ant terminal board: located on back cover (polystyrene)
	15P65173A57	COVER, cab back: incls ant terminal cover; less all other components
	1 5D68244A03	COVER, CRT: rear
	30S183A13	CORD, AC line
	15P65173A59	DOOR, control: incls ctrl overlay; less cloth panel overlay & hinge spring clips
	13P65173A55	ESCUTCHEON, control panel: UHF/ VHF/ incls slide ctrl overlay
	55C63659A28	GLIDE, dome: cab leg
	13E69943A02	GRILLE, speaker: less mtg spring & cloth panel overlay
	42B70093A01	HINGE, spring clip: control door (R & L)
	36D69994A01	KNOB, channel selector: UHF (outer)
	36D69994A02	KNOB, channel selector: VHF
	36D69982A02	KNOB, fine tune: UHF (inner)
	36D68218A35	KNOB, on-off vol & VHF fine tune
	36C69981A01	KNOB, slide: intensity & hue
	16B70115A01	LEG, cabinet
	13C69934A03	OVERLAY, control (on-vol VHF fine tune)
	13C69992A03	OVERLAY, control door (tone-bright-contrast-vertical)
	13D69957A17	OVERLAY, control door (cloth panel)
	13C69934A01	OVERLAY, nameplate (Motorola)
	13C69945A01	OVERLAY, slide control: intensity hue
	13D69957A18	OVERLAY, speaker grille (cloth panel)
	64A67158A02	PLATE, metal: leg mtg
	34D69996A04	SCALE, dial: UHF
	34D69996A03	SCALE, dial: VHF
	3K749169	SCREW, 8-15 x 5/8 (CRT frame, slide rails & back cover spring mtg)
	3S134013	SCREW, tpg: 8-15 x 1/2 (spkr grille retainer spring)
	3D66303A18	SCREW, tpg: 8-15 x 3/4 (CRT bezel to frame)
	3D66303A19	SCREW, tpg: 8-15 x 13/16 (interlock)
	3S136869	SCREW, tpg: 8-18 x 1-1/8 (phl flt blk ox) ctrl panel bezel to chassis
	50D67337A01	SPEAKER, 4 x 6 PM 16Ω VC
	41A70069A01	SPRING, speaker grille retainer

CABINET PARTS MODELS WU820FW, WU821FS & WU822FP

REF.	PART NO.	DESCRIPTION
	13P65173A60	BEZEL, control panel: incls ctrl door, ctrl esc & overlays; less ctrl door cloth panel overlay & spkr grille
	13E69927A02	BEZEL, CRT
	16E70196A01	CABINET, TV: console; walnut (WU820FW)
	16E70196A02	CABINET, TV: console; maple (WU821FS)
	16E70196A03	CABINET, TV console; pecan (WU822FP)
	42S10122A01	CLIP, secures line cord to ant cover on cab back
	42S10152A02	CLIP, push-on: spkr mtg
	42C67787A04	CLIP, spring: cab back mtg
	15D70134A01	COVER, ant terminal board: located on back cover (polystyrene)
	15P65173A57	COVER, cab back: incls ant terminal cover; less all other components
	15D68244A03	COVER, CRT: rear
	30S183A13	CORD, AC line
	15P65173A59	DOOR, control: incls ctrl overlay; less cloth panel overlay & hinge spring clips
	13P65173A55	ESCUTCHEON, control panel: UHF/ VHF/ incls slide ctrl overlay
	16E70196A05	GALLERY, cabinet (WU821FS)
	55C63659A28	GLIDE, dome: cab leg
	13E69943A02	GRILLE, speaker: less mtg spring & cloth panel overlay
	42B70093A01	HINGE, spring clip: control door (RH & LH)
	36D69994A01	KNOB, channel selector: UHF (outer)
	36D69994A02	KNOB, channel selector: VHF
	36D69982A02	KNOB, fine tune: UHF (inner)
	36D68218A35	KNOB, on-off vol & VHF fine tune
	36C69981A01	KNOB, slide: intensity & hue
	16B70197A01	LEG, cabinet (WU820FW)
	16B70198A01	LEG, cabinet (WU821FS)
	16B70199A01	LEG, cabinet (WU822FP)
	13C69934A03	OVERLAY, control (on-vol VHF fine tune)
	13C69992A03	OVERLAY, control door (tone-bright-contrast-vertical)

REF.	PART NO.	DESCRIPTION	REF.	PART NO.	DESCRIPTION
	34D69996A04	SCALE, dial: UHF	DIODES		
	34D69996A03	SCALE, dial: VHF			
	3K749169	SCREW, 8-15 x 5/8 (CRT frame, slide rails & back cover spring mtg)	D-1	48G10346A02	DIODE
			D-3	48S191A04	RECTIFIER, silicon (USE 48S191A07)
			D-4	48S137133	DIODE: D3A
	3S134013	SCREW, tpg: 8-15 x 1/2 (spkr grille retainer spring)	D-5	48G10346A01	DIODE (USE 48D67120A11)
			D-7	48G10346A01	DIODE (USE 48D67120A11)
	3D66303A18	SCREW, tpg: 8-15 x 3/4 (CRT bezel to frame)	D-8	48S137133	DIODE: D3A
			D-9	48G10346A02	DIODE
	3D66303A19	SCREW, tpg: 8-15 x 13/16 (interlock)	D-10	48S137133	DIODE: D3A
	3S136869	SCREW, tpg: 8-18 x 1-1/8 (phl flt blk ox) ctrl panel bezel to chassis	INTEGRATED CIRCUIT		
			IC-1	51M70177A01	COLOR DEMODULATOR
	50D67337A01	SPEAKER, 4 x 6 PM 16Ω VC	COILS & CHOKES		
	41A70069A01	SPRING, speaker grille retainer			
	13D69957A05	OVERLAY, control door (cloth panel - WU820FW)	L-1	24D69708A02	COMPENSATING: 47 uh
			L-2	24D68852A03	DELAY LINE
	13D69957A07	OVERLAY, control door (cloth panel - WU821FS)	L-3	24D69708A02	COMPENSATING: 47 uh
			L-4	24D68801A46	COMPENSATING: 800 uh
	13D69957A09	OVERLAY, control door (cloth panel - WU822FP)	L-5	24D68517A12	3.58 MHz OSCILLATOR
			L-6	24D68002A98	COMPENSATING: 5.6 uh
	13C69934A01	OVERLAY, nameplate (Motorola)	L-7	24D68517A21	HUE RANGE
	13C69945A01	OVERLAY, slide control: intensity hue	L-8	24D68801A50	COMPENSATING: 47 uh
			L-9	24D68801A48	COMPENSATING: 8.2 uh
	13D69957A06	OVERLAY, speaker grille (cloth panel - WU820FW)	L-10	24D68801A48	COMPENSATING: 8.2 uh
			L-11	24D68801A48	COMPENSATING: 8.2 uh
	13D69957A08	OVERLAY, speaker grille (cloth panel - WU821FS)	L-12	24D68801A03	COMPENSATING: 100 uh
			L-13	24D68801A02	COMPENSATING: 450 uh
	13D69957A10	OVERLAY, speaker grille (cloth panel - WU822FP)	TRANSISTORS		
	64A67158A02	PLATE, metal: leg mtg			

IF AUDIO PANEL "BA"

ELECTRICAL PARTS

REF.	PART NO.	DESCRIPTION	REF.	PART NO.	DESCRIPTION
D-1	48D67120A13	DIODE, low power (in panel BA-3 & later)	Q-1	48S134841	1ST COLOR IF: M4841
			Q-2	48S134841	2ND COLOR IF: M4841
	48D67120A02	DIODE, lower power (in panel BA-7 & later)	Q-3	48S134841	2ND VIDEO AMP: M4841
			Q-4	48S137002	VOLTAGE REGULATOR: A3M
Q-2	48S137171	TRANSISTOR: AGC amp; A6H (in panel BA-5 & later)	Q-5	48S134842	PULSE LIMITER & INVERTER (USE 48S134992)
			Q-6	48S137111	COLOR SYNC GATE & AMP: A5S
ELECTRICAL PARTS			Q-7	48S134842	CRYSTAL DRIVER: M4842 (USE 48S134992)
D-2	48D67120A02	DIODE, low power (in panel FA-3 & later)	Q-8	48S134841	CRYSTAL AMP: M4841
			Q-9	48S137115	ACC AMP: A5U

AFT PANEL "KA"

ELECTRICAL PARTS

			Q-10	48S137003	CRYSTAL OSCILLATOR: A3N
T-1	24D68501A24	AFT DISCRIMINATOR (in panel KA-1 & later	Q-11	48S137127	COLOR KILLER: P2S
			Q-12	48S134970	PHASE SPLITTER: A2T
			Q-13	48S134841	PHASE SHIFTER: M4841
			Q-14	48S134918	3.58 LIMITER: A1L

COLOR VIDEO PANEL "SA"

ELECTRICAL PARTS

			Q-15	48S137002	BLUE VIDEO OUTPUT: A3M
DIODES			Q-16	48S137002	GREEN VIDEO OUTPUT: A3M
D-9	48G10346A02	DIODE (in panel SA-9 & later)	Q-17	48S137002	RED VIDEO OUTPUT: A3M
D-10	48S137133	DIODE: D3A (in panel SA-10 & later)	Q-18	48S137127	SYNC & AGC TAKE-OFF: P2S

COILS & CHOKES			CONTROLS		
L-3	24D69708A02	COMPENSATING: 47 uh (in panel SA-13 & later)	R-76	18D66401A42	BLUE VIDEO DRIVE: 100Ω
			R-83	18D66401A41	GREEN VIDEO DRIVE: 100Ω
L-8	24D68801A50	COMPENSATING: 47 uh (in panel SA-16 & later)	R-89	18D66401A40	RED VIDEO DRIVE: 100Ω
L-13	24D68801A02	COMPENSATING: 450 uh (in panel SA-12 & later)	TRANSFORMERS		
			T-1	24D68517A18	2ND COLOR IF

COMPLETE COLOR VIDEO PANEL "CA" CODED CA16 THRU
CA49. Does not include "PA" Panel parts. See "PA" Panel Parts List

CRYSTALS					
Y-1	48C66865A04	3.58MHz CRYSTAL			

	1Y68662A59	COLOR VIDEO PANEL "CA" WITH "PA" PANEL: complete

MECHANICAL PARTS

	26B66745A05	HEAT SINK, for transistor Q4, 15, 16 & 17
	29S10134A29	LUG, connector: panel mtg
	43B68719A01	SPACER, transistor mtg: Q4, 15, 16 & 17

INSTAMATIC COLOR PRESET PANEL "PA" (PART OF "CA" PANEL)

ELECTRICAL PARTS

DIODES

D-1	48G10346A02	DIODE
D-2	48G10346A02	DIODE

ELECTRICAL PARTS

TRANSISTORS

Q-1	48S137172	1ST COLOR INT AMP: A6J

REF.	PART NO.	DESCRIPTION	REF.	PART NO.	DESCRIPTION
Q-2	48S137172	2ND COLOR INT AMP: A6J	**COILS & CHOKES**		
Q-3	48S137172	AND GATE: A6J			
Q-4	48S137172	AND GATE: A6J	**L-1Y**	24D66772A07	CHOKE, RF suppressor
Q-5	48S137021	DIODE, Zener: D1U	**L-2Y**	24D68801A01	COMPENSATING, 15 uh
			TRANSISTORS		
MECHANICAL PARTS					
	39S10184A31	CONTACT, card edge conn hsg	Q-1Y	48S134942	FUNCTION OUTPUT: A2B
	26B68924A02	COVER, coil shield: panel mtg; L5, 7 & T1-"CA" panel	Q-2Y	48S137014	1ST AUDIO AMP: A4A
			Q-3Y	48S137014	2ND AUDIO AMP: A4A
	15S10390A03	HOUSING, connector: card edge; 6 contact - less contacts	Q-4Y	48S134841	1ST HUE AMP: M4841
			Q-5Y	48S134841	2ND HUE AMP: M4841
			Q-6Y	48S134841	INTENSITY AMP: M4841
TRR-7 REMOTE RECEIVER			**CONTROLS**		
REMOTE PRE-AMP PANEL "U" (CODED U-0 THRU U-2)			R-62Y	18D66401A26	AUDIO CUT IN: 100K
ELECTRICAL PARTS			**TRANSFORMERS**		
TRANSISTORS					
			T-1Y	24D69051A03	VOLUME DOWN
Q-1U	48S137014	1ST PRE-AMP (A4A)	T-2Y	24D69051A03	VOLUME UP
Q-2U	48S134842	2ND PRE-AMP (M4842 - USE 48S134992)	T-3Y	24D69051A03	HUE TO RED
	48S134841	2ND PRE-AMP (IN PANEL U-2 & LATER)	T-4Y	24D69051A03	HUE TO GREEN
Q-3U	48S134942	3RD PRE-AMP (A2B)	T-5Y	24D69051A03	INTENSITY DOWN
			T-6Y	24D69051A03	INTENSITY UP
CONTROL					
R-15U	18D66401A33	PRE-AMP SENSITIVITY: 2K	**CONTROL VOLTAGE SWITCH PANEL (PART OF COMPLETE "Y" PANEL)**		
			TRANSISTOR		
REMOTE POWER SUPPLY PANEL "X" (CODED X-0 THRU X-7)					
ELECTRICAL PARTS			Q-7	48S134841	HUE OUTPUT: M4841
			SWITCH		
MISCELLANEOUS ELECTRICAL PARTS			SW-1Y	40D69977A02	SLIDE SWITCH
E-1X	80D69349A01	RELAY, channel change	**MECHANICAL PARTS**		
E-2X	80D69349A02	RELAY, on-off			
E-3X	48C65837A02	DIODE, crystal		29S10134A29	CONNECTOR, recept: module panel mtg
E-4X	48S191A05	RECTIFIER, silicon		15S10183A16	CONNECTOR, recept: 1 contact; less contact (hue)
E-5X	48S191A05	DIODE			
E-6X	48S191A05	DIODE		39S10184A01	CONTACT, recept: for connector 15S10183A16
E-7X	48S137034	DIODE, silicon: Zener (D1W)			
E-8X	65S10081A06	LAMP, neon: 3AD			
E-9X	65S10081A06	LAMP, neon: 3AD	**MECHANICAL PARTS FOR PANELS "U", "Y", & "X"**		
E-10X	65S10081A06	LAMP, neon: 3AD		42S10152A07	CLIP, screw retainer: metal; panel mtg to brkt
E-11X	65S10081A06	LAMP, neon: 3AD		31C68421B02	CONNECTOR, panel: 5 pin; panel mtg
TRANSISTORS				15S10183A20	CONNECTOR, plug: 12 contact; less contact (natural)
Q-1X	48S134942	ALL FUNCTION DRIVER (A2B)		15S10183A28	CONNECTOR, plug: 12 contact; less contact (red)
Q-2X	48S134841	CHANNEL CHANGE RELAY DRIVER (M4841)		39S10184A02	CONTACT, plug: for 15S10183A20 & A28 connectors
Q-3X	48S134933	ON/OFF RELAY AMPLIFIER (A1V)		5B69911A01	GROMMET, plastic: panel mtg
Q-4X	48S137032	ON/OFF RELAY DRIVER (P2E)		9C67349B03	RECEPTACLE, phono plug: on pre-amp panel "U"
Q-5X	48S134903	REGULATOR (A1F)		14C68842A01	SHIELD, plastic: covers metal PC mtg clips (42S10152A07)
TRANSFORMERS				26A66745A01	SINK, heat: transistor Q5X: on power panel "X"
T-1X	24D69051A01	DRIVER		43B68719A01	SPACER, transistor: Q5X; on power panel "X"
T-2X	24D69015A02	CHANNEL CHANGE		41B69399A01	SPRING, grounding: control panel "Y" (2 used)
REMOTE CONTROL PANEL PANEL "Y" (CODED Y-0 THRU Y-9)					
	1U68674A02	REMOTE CONTROL PANEL "Y": complete	**TRT-6 TRANSMITTER**		
ELECTRICAL PARTS			**ELECTRICAL PARTS**		
			CAPACITORS		
MISCELLANEOUS ELECTRICAL PARTS			C-13 thru		
D-28Y	48D67120A11	DIODE (in panel Y-9 & later)	C-19	20B90199A01	20 pf trimmer (7 req'd)
D-29Y	48D67120A11	DIODE (in panel Y-9 & later)			
E-1Y	48D67120A06	DIODE (USE 48P65145A74)	E-1	50P65149A87	SPEAKER, incls mtg brkt
E-2Y	48D67120A06	DIODE (USE 48P65145A74)			
E-3Y	48S137074	DIODE: D2J	L-1	24B90201A01	OSCILLATOR COIL
E-4Y	48S137074	DIODE: D2J			
E-5Y	48D67120A06	DIODE	Q-1	48S134458	TRANSISTOR: 2SB56
E-6Y	48D67120A06	DIODE			
E-7Y	48S137074	DIODE: D2J	R-1	6S127805	15K 10% 1/4W
E-8Y	48S137074	DIODE: D2J			
E-9Y	48D67120A06	DIODE (USE 48P65145A74)			
E-10Y	48D67120A06	DIODE (USE 48P65145A74)			
E-11Y	48S137074	DIODE: D2J			
E-12Y	48S137074	DIODE: D2J			
E-19Y	51D69344A01	MODULE, audio memory			
E-20Y	51D69344A01	MODULE, hue memory			
E-21Y	51D69344A01	MODULE, intensity memory			

REF.	PART NO.	DESCRIPTION	REF.	PART NO.	DESCRIPTION
				24V68639A80	COIL, CRT degausser: complete (WT901, WU911, 912, 913, 914, 937, 938 & 939)
MECHANICAL PARTS			**CONTROLS**		
	84P65149A88	BOARD, printed circuit: incls contact springs	R-100	18D67858A07	AGC DELAY, 1KΩ (ATS, STS-934)
	7B90188A01	BRACKET, trimmer capacitor mtg	R-104 A&B	18D68721A08	CONTRAST & BLUE G2, contrast 750K: blue G2 3 meg (ATS, STS-934)
	38B90203A01	BUTTON, push: channel	R-108	18D67858A10	MASTER BRIGHTNESS, 3.5K (ATS, STS-934)
	38B90202A01	BUTTON, push: hue, intensity, volume	R-109	18D67858B15	PRE-SET CONTRAST, 750Ω (STS-934)
	15D90182A01	CASE, bottom	R-206 A&B	18D68221A09	BRIGHTNESS & RED G2, brightness 1KΩ: red G2 3 meg (ATS, STS-934)
	15P65149A85	CASE, top: incls control overlay	R-215	18D67502A20	FOCUS, 10 meg (ATS, STS-934)
	15B90183A01	ESCUTCHEON, speaker	R-219	18D67858B13	PRE-SET BRIGHTNESS, 5K (STS-934)
	5B90197A01	EYELET (contact spring mtg to PC board)	R-300 A&B	18D68221A11	TONE & VIDEO PEAKING, tone 5K: vid peak 3.5K (ATS, STS-934)
	2B90192A01	NUT, button guide assembly to PC board – 3 req'd	R-301	18D68021A15	VOLUME, 50KΩ: incls (SW800) on-off switch (STS-934)
	64B90185A01	PLATE, button guide: plastic	R-603 A&B	18D68221A10	VERTICAL HOLD & GRN G2, vert hold 1 meg: grn G2 3 meg (ATS, STS-934)
	64B90186A01	PLATE, guide: fibre; button spring guide	R-900	18D68443A17	INTENSITY, 5K (STS-934)
	3B90195A01	SCREW, mach: 1/4" phl flat (bottom case mtg - 2 req'd)	R-901	18D68443A16	HUE, 25K (STS-934)
	3B90193A01	SCREW, mach: 1" phl flat (button guide assembly mtg – 3 req'd)	R-902	18D67858B15	PRE-SET INTENSITY, 750Ω (STS-934)
	3B90194A01	SCREW, tpg: 5/16" phl rnd (PC board mtg - 2 req'd)	R-903	18D67858B14	PRE-SET HUE, 25K (STS-934)
	43B90207A01	SPACER, button guide assembly mtg – 3 req'd	**TRANSFORMERS**		
	41B90206A01	SPRING, button return	T-4X	25D68865A01	POWER (remote power panel – ATS-934)
	41P65149A86	SPRING, contact: incls contacts			
	46B90191A01	STUD, threaded (PC board mtg & bottom cover mtg)	**MECHANICAL PARTS**		
	29B90187A01	TERMINAL, battery		7D70099A01	BRACKET, CRT mtg: less cloth tape (WT677GW)
	80C69015C01	TRANSMITTER, TRT-6: complete		7C70405A05	BRACKET, CRT mtg: less rubber sleeve (WT680, WT682 & TT683)
	4B90196A01	WASHER (button guide plate mtg – 3 req'd)		7D68564A09	BRACKET, CRT mtg: less rubber sleeve (WU834, 835, & 836)

UNIQUE ELECTRICAL & MECHANICAL PARTS (NOT ON PANELS FOR ATS & STS-934 CHASSIS.

REF.	PART NO.	DESCRIPTION	REF.	PART NO.	DESCRIPTION
ELECTRICAL PARTS				7D70236A10	BRACKET, CRT mtg: less rubber sleeve (WT901, WU911, 912, 913 & 914)
MISCELLANEOUS ELECTRICAL PARTS				7D70236A01	BRACKET, CRT mtg: less rubber sleeve (WU937, 938 & 939)
	30V68644A74	MICROPHONE, incls cable & plug (ATS-934)		42C69663A01	CLAMP, degausser coil mtg (1-3/4" long - loop on one end - WT677, WU834, 835, 836, WT901, WU911, 912, 913, 914, 937, 938 & 939)
C-13U	21S120936	560 pf 20% 500V Z5F cer disc (ATS-934 - USE 21R131336)		42C69663A02	CLAMP, degausser coil mtg (3" long - loop on both ends - WT677GW, WU934, 835 & 836)
C-101	21S180A98	100 pf 10% 500V Z5F (934-A23 & later)		42C69663A03	CLAMP, degausser coil mtg (3-1/8" long - loop on one end - WT677GW)
C-205	23D65808A37	10 mf 50V lytic (934A-16 & later)		42C65572A18	CLAMP, degausser coil mtg (snap in CRT shield mtg type - WT680, WT682, TT683, WT901, WU911, 912, 913, 914, 937, 938, & 939)
C-206	21S180C56	.001 mf +80-20% 500V Z5U (934-A26 & later)	P-9	15S10183A10	CONNECTOR, plug: 8 contact; less contacts
C-805	23S10255A35	1000 mf 25V lytic (934-A24 & later)	P-10	15S10183A10	CONNECTOR, plug: 8 contact; less contacts (remote - ATS-934)
E-22Y	1D69162A12	MOTOR & GEAR BOX, complete (ATS-934)	P-11	15S10183A10	CONNECTOR, plug: 8 contact; less contacts (VHF tuner - ATS-934)
	1D69162A10	ARMATURE, motor: incls nylon gear (ATS-934)	P-12	15S10183A28	CONNECTOR, plug: 12 contact; less contacts (remote-red - ATS-934)
	1D69162A14	GEAR BOX, less motor (ATS-934)	JL-1	15S10183A19	CONNECTOR, receptacle: 12 contact; less contacts (remote-wht - ATS-934)
E-23Y A,B,C	1D69162A13	SWITCH, leaf: on gear box motor (ATS-934)	JL-1A	15S10183A09	CONNECTOR, receptacle: 8 contact; less contacts (remote - ATS-934)
E-25Y	40D69358A05	SWITCH, leaf: volume; incls pushbutton (ATS-934)	JL-2	15S10183A27	CONNECTOR, receptacle: 12 contact; less contacts (remote-red - ATS-934)
E-26Y	40D69358A05	SWITCH, leaf: hue; incls pushbutton (ATS-934)	JL-2A	15S10183A27	CONNECTOR, receptacle: 12 contact; less contacts (remote-red - ATS-934)
E-27Y	40D69358A05	SWITCH, leaf: intensity; incls pushbutton (ATS-934)	S-9	15S10183A09	CONNECTOR, receptacle: 8 contact; less contacts (motor - ATS-934)
PL-802	65S135685	LAMP, min incand: 6.3V - .15A #755 (pre-set STS-934)	S-11	15S10183A09	CONNECTOR, receptacle: 8 contact; less contacts (VHF tuner -ATS-934)
PL-803	65S10081A08	LAMP, glow: power on-off indicator #C2A (ATS-934)		39S10184A02	CONTACT, plug: for connectors (P9, 10, 11 & 12 - ATS-934)
SW-800	40D66846A06	SWITCH, power on-off (ATS-934)		39S10184A01	CONTACT, recept: for connectors (JL1, 1A, 2, 2A & S9, S11 - ATS-934)
SW-901	40V68645A16	SWITCH, inst. color pre-set: incls mtg brkt (STS-934)			
COILS & CHOKES					
L-100	24D68002A28	COMPENSATING, 100 uh (934-A23 & later)			
L-701	24G10250A10	DEFLECTION YOKE, 92°			
L-801	24V68636A27	COIL, CRT degausser: complete (WT677GW)			
	24V68645A49	COIL, CRT degausser: complete (WT680, WT682, & TT683)			
	24V68636A98	COIL, CRT degausser: complete (WU834. 835 & 836)			

REF.	PART NO.	DESCRIPTION	REF.	PART NO.	DESCRIPTION
	5S10115A19	GROMMET, rubber: power on-off indicator lamp; PL803 (ATS-934)		13C69992A03	OVERLAY, control door: "tone-bright -contrast-vertical"
	29K580544	PLUG, 75Ω antenna input		13C69957A03	OVERLAY, control door (cloth panel)
	3B68012A04	SCREW, mch: 1/4-20 x 5/8; CRT mtg		13C69934A17	OVERLAY, nameplate (Motorola)
	43C70407A02	SLEEVE, rubber: CRT mtg brkt (WT680, WT682 & TT683)		13C69945A01	OVERLAY, slide control (intensity/hue)
	43C70407A04	SLEEVE, rubber: CRT mtg brkt (WU834, 835 & 836)		13C69957A04	OVERLAY, speaker grille (cloth panel)
	43C70407A03	SLEEVE, rubber: CRT mtg brkt (WT901, WU911, 912, 913, 914, 937, 938 & 939)		34D69996A04	SCALE, dial: UHF
	9S10143A17	SOCKET, pre-set lamp (STS-934)		34D69996A03	SCALE, dial: VHF
	41D65987A25	SPRING, CRT aquadag grnd (screw mtg type)		3S138218	SCREW, tpg: 8 x 5/8; VHF ant mtg
	41D65987A27	SPRING, CRT aquadag grnd (snap in CRT shield type)		3S136869	SCREW, tpg: 8-18 x 1-1/8 (phl flt blk ox) ctrl panel bezel to chassis
	41D69152A01	SPRING, CRT shield mtg (WU834, 835 & 836)		3D66303A18	SCREW, tpg: 8-15 x 3/4; CRT bezel to cab
	41A70031A03	SPRING, CRT shield mtg (WT901, WU911, 912, 913, 914, 937, 938 & 939)		3D66303A19	SCREW, tpg: 8-15 x 13/16; inter-lock
	42C65572A16	STRAP, adjustable: degausser coil mtg (WT677GW)		50D68384A02	SPEAKER, 4" PM 16Ω VC
	42D67027A04	STRAP, metal: CRT mtg (WT677GW)		41B65987B09	SPRING, CRT bezel ground
	42D67027A10	STRAP, metal: CRT mtg (WT680, WT682 & TT683)		41A70069A01	SPRING, speaker grille retainer
	42D67027A11	STRAP, metal: CRT mtg (WU834, 835 & 836)	**MODEL WT680GWA**		
	42D67027A09	STRAP, metal: CRT mtg (WT901, WU911, 912, 913, 914, 937, 938 & 939)		85D67625A08	ANTENNA, UHF: "bow tie"

UNIQUE ELECTRICAL & MECHANICAL PARTS FOR AOPTT-429 VHF TUNER. For complete replacement parts list, refer to OPTT-429 VHF TUNER in CTV5 (TS-934) Service Manaual (68P65155A59)

REF.	PART NO.	DESCRIPTION	REF.	PART NO.	DESCRIPTION
E-24Y	40P65174A23	SWITCH, channel stop skip		85D68375A03	ANTENNA, VHF: dipole; complete
	44X90277A07	GEAR, fine tune: incls bushing, clutch & metal gear		13P65175A34	BEZEL, control panel: incls ctrl door, pre-set lens & all overlays; less speaker grille
	47X90276A14	SHAFT, channel selector: incls detent wheel & pre set screw, coil & osc stator assembly		13E70323A02	BEZEL, CRT
				75B67953A02	BUMPER, cab foot
	77V68640A48	VHF TUNER, AOPTT-429: complete; incls plugs, convertor cable, mtg brkt, and all other out-board component parts; less gear motor assembly		16E70084A11	CABINET, TV: metal; walnut (incls chassis slide rail)
				42S10122A01	CLIP, secures line cord to cab back
				42S10152A02	CLIP, push on; spkr mtg
	CABINET PARTS			42C67787A04	CLIP, spring: cab back mtg
				30S183A13	CORD, AC line
MODEL WT677GW				15E70132D01	COVER, cabinet back: less all other components
	85D67625A08	ANTENNA, UHF: "bow tie"		15E68244A02	COVER, CRT: rear
	85D68375A03	ANTENNA, VHF: dipole; complete		15P65175A33	DOOR, control: incls ctrl over-lay; less hinge spring clips
	13P65175A32	BEZEL, control panel: incls ctrl door, ctrl esc & overlays; less speaker grille		13E70068A08	GRILLE, speaker: less mtg spring
	13E70019A02	BEZEL, CRT		42B70093A01	HINGE, spring clip: ctrl door (RH & LH)
	75B67953A02	BUMPER, cabinet foot		36D70324A04	KNOB, channel selector: UHF (outer)
	16E70084A07	CABINET, TV: metal; walnut (incls chassis slide rails)		36D70324A06	KNOB, channel selector: VHF
	42S10122A01	CLIP, secures line cord to cab back		36D70325A04	KNOB, fine tune: UHF (inner)
	42S10152A02	CLIP, push-on: spkr mtg		36D69941A06	KNOB, on-off vol & VHF fine tune
	42C67787A04	CLIP, spring: cab back mtg		36C69981A01	KNOB, slide: intensity & hue
	30S183A13	CORD, AC line		61A70527A01	LENS, color pre-set
	15E70132D01	COVER, cabinet back: less all other components		13C69934A14	OVERLAY, control (on vol VHF fine tune)
	15D68244A02	COVER, CRT: rear (USE 15D67156A02)		13C69992A05	OVERLAY, control door (tone-vid peak-bright, etc.)
	15P65173A54	DOOR, control: incls cloth panel & ctrl overlays; less hinge spring clips		13C69993A03	OVERLAY, control panel: UHF/VHF slide ctrl; less pre-set lens
	13P65173A55	ESCUTCHEON, control panel: UHF/VHF; incls slide control overlay		13C69934A13	OVERLAY, nameplate ("Quasar II)
	13P65173A53	GRILLE, speaker: incls cloth panel overlay; less mtg spring		36D70491A02	PUSHBUTTON, color pre-set
	42B70093A01	HINGE, spring clip: ctrl door (RH & LH)		34D69996A02	SCALE, dial: UHF
	36D69994A01	KNOB, channel selector: UHF (outer)		34D69996A01	SCALE, dial: VHF
	36D69994A02	KNOB, channel selector: VHF		3S136869	SCREW, tpg: 8-18 x 1-1/8 (phl flt blk ox) ctrl panel bezel to chassis
	36D69982A01	KNOB, fine tune: UHF (inner)		3D66303A18	SCREW, tpg: 8-15 x 3/4; CRT bezel to cab
	36D68218A35	KNOB, on-off vol & VHF fine tune		3D66303A19	SCREW, tpg: 8-15 x 13/16; inter-lock
	36C69981A01	KNOB, slide: intensity & hue		50D68384A02	SPEAKER, 4" PM 16Ω VC
	13C69934A15	OVERLAY, control: "on vol/VHF fine tune"		41B65987B09	SPRING, CRT bezel grnd
				41A70069A02	SPRING, spkr grille retainer
			MODEL WT682GWA		
				85D67625A08	ANTENNA, UHF: "bow-tie"
				85D68375A03	ANTENNA, VHF: dipole; complete
				13P65174A58	BEZEL, control panel: incls ctrl door, ctrl esc & overlays; less speaker grille
				13E70323B01	BEZEL, CRT
				75B67953A02	BUMPER, cab foot
				16E70084A11	CABINET, TV: metal; walnut (incls chassis slide rail)

REF.	PART NO.	DESCRIPTION
	42S10122A01	CLIP, secures line cord to cab back
	42S10152A02	CLIP, push-on: spkr mtg
	42C67787A04	CLIP, spring: cab back mtg
	30S183A13	CORD, AC line
	15E70132D01	COVER, cabinet back: less all other components
	15E68244A02	COVER, CRT: rear (USE 15D67156A02)
	15P65174A60	DOOR, control: incls ctrl overlay; less hinge spring clips
	13P65174A59	ESCUTCHEON, control panel: UHF/VHF incls slide ctrl overlay & pre-set indicator jewel
	13E70068A05	GRILLE, speaker: less mtg spring
	42B70093A01	HINGE, spring clip: ctrl door (RH & LH)
	36D70324A01	KNOB, channel selector: UHF (outer)
	36D70324A03	KNOB, channel selector: VHF
	36D70325A02	KNOB, fine tune: UHF (inner)
	36D69941A03	KNOB, on-off vol & VHF fine tune
	36C69981A02	KNOB, slide: intensity & hue
	13C69934B08	OVERLAY, control (on-vol VHF fine tune)
	13C69992A05	OVERLAY, control door (tone-vid peak-bright, etc.)
	13C69934B10	OVERLAY, nameplate ("Quasar II")
	13C69945A08	OVERLAY, slide control (intensity & hue)
	36D70491A06	PUSHBUTTON, color pre-set
	34D69996A05	SCALE, dial: UHF
	34D69996A06	SCALE, dial: VHF
	3S136869	SCREW, tpg: 8-18 x 1-1/8 (phl flt blk ox) ctrl panel bezel mtg
	3D66303A18	SCREW, tpg: 8-15 x 3/4; CRT bezel to cab
	3D66303A19	SCREW, tpg: 8-15 x 13/16; interlock
	50D68384A02	SPEAKER, 4" PM 16Ω VC
	41B65987B09	SPRING, CRT bezel grnd
	41A70069A02	SPRING, spkr grille retainer

MODEL TT683GW

REF.	PART NO.	DESCRIPTION
	85D67625A08	ANTENNA, UHF: "bow tie"
	85D68375A03	ANTENNA, VHF: dipole; complete
	13P65175A39	BEZEL, control panel: incls ctrl door, ctrl esc & all overlays; less speaker grille
	13E70323B01	BEZEL, CRT
	75B67953A02	BUMPER, cabinet foot
	16E70084A11	CABINET, TV: metal; walnut (incls chassis slide rail)
	42S10122A01	CLIP, secures line cord to cab back
	42S10152A02	CLIP, push on: spkr mtg
	42C67787A04	CLIP, spring: cab back mtg
	30S183A13	CORD, AC line
	15D70132E01	COVER, cabinet back: less all other components
	15E68244A02	COVER, CRT: rear (USE 15D67156A02)
	15P65174A60	DOOR, control: incls ctrl overlay less hinge spring clips
	13P65175A40	ESCUTCHEON, control panel: UHF/VHF; incls ctrl overlay "vol-intensity-hue" & indicator jewel
	13P65175A38	GRILLE, speaker: incls decorative mike ring; less mtg spring
	15C69376A02	HOLDER, remote transmitter
	42B70093A01	HINGE, spring clip: ctrl door (RH & LH)
	36D70324A01	KNOB, channel selector: UHF (outer)
	36D70324A03	KNOB, channel selector: VHF
	36D70325A02	KNOB, fine tune: UHF (inner)
	36D69941A03	KNOB, on-off, vol & VHF fine tune
	13C69934A12	OVERLAY, control (master power - VHF fine tune)
	13D69945B09	OVERLAY, control (vol-intensity-hue)
	13C69992A05	OVERLAY, control door (tone-vid peak-bright, etc.)
	13C69934B10	OVERLAY, nameplate (Quasar II)
	34D69996A05	SCALE, dial: UHF
	34D69996A06	SCALE, dial: VHF
	3S136869	SCREW, tpg: 8-18 x 1-1/8 (phl flt blk ox) ctrl panel bezel mtg
	3D66303A18	SCREW, tpg: 8-15 x 3/4; CRT bezel to cab
	3D66303A19	SCREW, tpg: 8-15 x 13/16; interlock
	50D68384A02	SPEAKER, 4" PM 16Ω VC
	41B65987B09	SPRING, CRT bezel grnd
	41A70069A02	SPRING, spkr grille retainer

MODELS WU834GWA, WU835GSA, WU836GPA

REF.	PART NO.	DESCRIPTION
	13P65175A35	BEZEL, control panel: incls ctrl door, pre-set lens & all overlays; less speaker grille
	13E69927A05	BEZEL, CRT
	16E70255A01	CABINET, TV: console; walnut (WU834GWA)
	16E70533A01	CABINET, TV: console; maple (WU835GSA)
	16E70533A02	CABINET, TV: console; pecan (WU836GPA)
	55C63659A23	CASTER, less grip neck socket (WU835GSA & WU836GPA)
	42S10122A01	CLIP, secures line cord to cab back
	42S10152A02	CLIP, push on: spkr mtg
	42C67787A04	CLIP, spring: cab back mtg
	30S183A13	CORD, AC line
	15D70134D01	COVER, ant terminal board: located on back cover (polyestyrene)
	15P65173A57	COVER, cabinet back: incls ant terminal cover; less all other components
	15E68244A03	COVER, CRT: rear
	15P65175A36	DOOR, control: incls ctrl overlay; less hinge spring clips
	16E70196A05	GALLERY, cabinet (WU835GSA)
	13E69943A05	GRILLE, speaker: less mtg spring
	42B70093A01	HINGE, spring clip: ctrl door (RL & LH)
	36D70324A04	KNOB, channel selector: UHF (outer)
	36D70324A06	KNOB, channel selector: VHF
	36D70325A04	KNOB, fine tune: UHF (inner)
	36D69941A06	KNOB, on-off vol & VHF fine tune
	36C69981A01	KNOB slide: intensity & hue
	16E70115A01	LEG, cabinet (WU834GWA)
	61A70527A01	LENS, color pre-set
	13C69934A14	OVERLAY, control (on vol VHF fine tune)
	13C69992A05	OVERLAY, control door (tone-vid peak-bright, etc.)
	13C69993A03	OVERLAY, control panel: UHF/VHF slide ctrl; less pre-set lens
	13C69934A13	OVERLAY, nameplate ("Quasar II")
	64A67158A02	PLATE, metal: leg mtg (WU834GWA)
	36D70491A02	PUSHBUTTON, color pre-set
	34D69996A02	SCALE, dial: UHF
	34D69996A01	SCALE, dial: VHF
	3S136869	SCREW, tpg: 8-18 x 1-1/8 (phl flt blk ox) ctrl panel bezel mtg
	3D66303A18	SCREW, tpg: 8-15 x 3/4; CRT bezel to cab back
	3D66303A19	SCREW, tpg: 8-15 x 13/16; interlock
	55C63659A24	SOCKET, grip neck: for caster (WU835GSA & WU836GPA)
	50D67337A01	SPEAKER, 4 x 6 PM 16Ω VC
	41A70069A02	SPRING, spkr grille retainer

MODELS WT901GWA, WU911GWA, WU912GWA, WU913GWA, WU914GPA

REF.	PART NO.	DESCRIPTION
	13P65175A35	BEZEL, control panel: incls ctrl door, pre-set lens & all overlays; less speaker grille
	13E70276A02	BEZEL, CRT
	75B67953A02	BUMPER, cabinet foot (WT901GWA)
	16E70462A01	CABINET, TV: metal; dk brn (incls chassis slide rail - WT901GWA)
	16E70514A01	CABINET, TV: console; walnut (WU911GWA)
	16E70394A01	CABINET, TV: console; walnut (WU912GWA)
	16E70515A01	CABINET, TV: console; walnut (WU913GWA)
	16E70515A02	CABINET, TV: console; pecan (WU914GPA)
	55C63659A23	CASTER, less grip neck socket (WU913GWA & WU914GPA)

REF.	PART NO.	DESCRIPTION
COILS		
L-1L	24D68801A06	COMPENSATING: 18 uh
L-2L	24D68801A13	COMPENSATING: 20 uh
L-3L	24D68801A25	COMPENSATING: 100 uh; incls 1.8K resistor
L-4L	24D68801A25	COMPENSATING: 100 uh; incls 1.8K resistor
L-5L	24D68801A25	COMPENSATING: 100 uh; incls 1.8K resistor
TRANSISTORS		
Q-1L	1V68611A47	RED DRIVE: incls heat sink (P1N)
Q-2L	1V68611A47	BLUE DRIVE: incls heat sink (P1N)
Q-3L	1V68611A47	GREEN DRIVE: incls heat sink (P1N)
Q-4L	48S134953	BRIGHTNESS CONTROL (A2K)
Q-5L	48S134953	ABL DRIVER (A2K)
Q-6L	48S134842	CRT BLANKER DRIVER (4842)
Q-7L	48S134933	CRT BLANKER OUTPUT (A1V)
TRANSFORMERS		
T-1L	24D68517A10	3.58Mc TRAP
T-2L	24D68517A10	3.58Mc TRAP
T-3L	24D68517A10	3.58Mc TRAP
		MECHANICAL PARTS
	29S10134A20	CONNECTOR, recept: PLAcir chassis mtg (15 used)

VIDEO OUTPUT PANEL "M"

ELECTRICAL PARTS

REF.	PART NO.	DESCRIPTION
MISCELLANEOUS ELECTRICAL PARTS		
E-1M	80C68147A02	SPARK GAP
E-2M	80C68147A02	SPARK GAP
E-3M	80C68147A02	SPARK GAP
COILS & CHOKES		
L-1M	24D66772A13	CHOKE, horiz suppressor
L-4M	24D66772A13	CHOKE, horiz suppressor
L-6M	24D66772A13	CHOKE, horiz suppressor
CONTROLS		
R-4M	18D68764A01	VIDEO DRIVE (RED): 10K
R-7M	18D68764A01	VIDEO DRIVE (BLUE): 10K
R-11M	18D68764A02	VIDEO PEAKING: 1K
R-12M	18D68764A01	VIDEO DRIVE (GREEN): 10K
TRANSISTORS		
Q-1M	48S134927	RED OUTPUT A1S: use A3M
	48S137002	RED OUTPUT A3M
Q-2M	48S134927	BLUE OUTPUT A1S: use A3M
	48S137002	BLUE OUTPUT A3M
Q-3M	48S134927	GREEN OUTPUT A1S: use A3M
	48S137002	GREEN OUTPUT A3M
		MECHANICAL PARTS
	7D68730A01	BRACKET, PLAcir chassis mtg
	42S10152A07	CLIP, metal: PLAcir chassis mtg (2 used)
	31C68421A01	CONNECTOR, panel: 5-pin PLAcir chassis mtg
	29S10134A20	CONNECTOR, recept (5 used)
	39S10184A09	CONTACT, plug (4 used)
	5B68518A01	GROMMET, plastic: PLAcir chassis mtg
	14C68986B01	INSULATOR, paper: covers video output panels (23TS-915 only)
	14D69245A01	INSULATOR: covers video output panel (TS-919 only)
	14C66650A01	INSULATOR, paper: for shafts (2 used)
	47C68765A01	SHAFT, plug-in: white
	47C68765A02	SHAFT, plug-in: red
	47C68765A03	SHAFT, plug-in: green
	47C68765A04	SHAFT, plug-in: blue
	26A66745A01	SINK, heat: transistor
	43B68719A01	SPACER, transistor mtg

REF.	PART NO.	DESCRIPTION
	1V68612A29	STRAP, braded: grnd PC panel; incls terminal & recept

FINE TUNING INDICATOR PANEL "P"

ELECTRICAL PARTS

REF.	PART NO.	DESCRIPTION
MISCELLANEOUS ELECTRICAL PARTS		
E-1P	48C65837A02	DIODE, crystal
E-2P	48C65837A02	DIODE, crystal
COILS		
L-1P	24D68501A09	45.75Mc COIL
L-2P	24D66772A12	CHOKE, 7.5 uh in panel P-6
L-3P	24D66772A12	CHOKE, 7.5 uh in panel P-6
TRANSISTORS		
Q-1P	48S134933	1ST DRIVER (A1V)
Q-2P	48S134910	2ND DRIVER (P1C)
Q-3P	48S134941	OUTPUT (A2A)
CONTROLS		
R-3P	18D66401A28	FTI ADJ 250K
		MECHANICAL PARTS
	15S10183A15	CONNECTOR, plug: less contact (blue lead)
	29S10134A15	CONNECTOR, recept (green lead)
	29C10134A11	CONNECTOR, recept (red lead)
	39C10184A02	CONTACT: for 15S10183A15 (blue lead)

FINE TUNING LOCK PANEL "T"

ELECTRICAL PARTS

REF.	PART NO.	DESCRIPTION
MISCELLANEOUS ELECTRICAL PARTS		
E-1T	48C65837A02	DIODE, crystal
E-2T	48C65837A02	DIODE, crystal
COILS & CHOKES		
L-1T	24D68501A09	COIL, 45.75 lite ind
L-2T	24D66772A12	CHOKE, IF resonant
L-3T	24D66772A12	CHOKE, IF resonant
TRANSISTORS		
Q-1T	48S134937	FTL AMP: A1Z
Q-2T	48S134937	FTL OUTPUT: A1Z
Q-3T	48S134937	FTL DETECTOR: A1Z
	48S137033	FTL DETECTOR: A4G
Q-4T	48S134941	FTL OUTPUT: A2A
CONTROL		
R-12T	18D66401A30	FTL ADJ: 50K
TRANSFORMERS		
T-1T	24D68517A11	DISCRIMINATOR
	15S10183A15	CONNECTOR, plug: less contacts (blue & gray lead)
	39S10184A02	CONTACT: for 15S10183A15
	29S10134A15	LUG, terminal (grn lead)
	29S10134A11	LUG, terminal (red lead)

COLOR PANEL "S"

ELECTRICAL PARTS

REF.	PART NO.	DESCRIPTION
MISCELLANEOUS ELECTRICAL PARTS		
E-1S	48S66865A04	CRYSTAL, 3.58Mc
E-2S	48C65837A02	CRYSTAL, diode
E-3S	48D67120A05	DIODE, low power

REF.	PART NO.	DESCRIPTION
PL-801	65S135685	LAMP, min incand: 6.3V #755; VHF (E16 & F18TS-929)
PL-802	65S135685	LAMP, min incand: 6.3V #755; pre-set (E16 &F18TS-929)
SW-7	1V68639A38	SWITCH, slide: AFT; incls mtg brkt (E16TS-929)
	1V68635A63	SWITCH, slide: AFT; incls mtg brkt (F18TS-929)
SW-901	1V68646A49	SWITCH, instamatic color (pre-set) incls mtg brkt (E16TS-929)
	1V68646A63	SWITCH, instamatic color (pre-set) incls mtg brkt (F18TS-929)

CONTROLS FOR E16TS & F18TS-929 ONLY.

REF.	PART NO.	DESCRIPTION
R-109	18D67858A19	PRE SET CONTRAST, 750Ω: (E16 & F18TS-929)
R-219	18D67858A17	PRE SET BRIGHTNESS, 5K (E16 & F18TS-929)
R-902	18D67858A19	PRE SET INTENSITY, 750Ω (E16 & F18TS-929)
R-903	18D67858A18	PRE SET HUE, 25K (E16 & F18TS-929)

MECHANICAL PARTS (NOT ON PANELS)

REF.	PART NO.	DESCRIPTION
	1V68639A04	ANTENNA BOARD, incls ant terminals, terminal strip & VHF ant shielded cable: less balun coil & res cap (16 & E16TS-929)
	1V68635A32	ANTENNA BOARD, incls ant terminals, terminal strips, nylon grommet, VHF ant shielded cable: less balun coil & res cap (18 & F18TS-929)
	7A70002A01	BRACKET, CRT mtg: on cab front (16 &E16TS-929)
	7B70033A01	BRACKET, CRT mtg: on cab front (18 &F18TS-929)
	75C66381A20	BUMPER, rubber: CRT spacer (16 & E16TS-929)
	30V68640A81	CABLE, shielded: incls plug & clamp; ant to VHF tuner (16 & E16TS-929)
	30V68635A33	CABLE, shielded: incls plug & clamp; ant to VHF tuner (18 & F18TS-929)
	42D65572A18	CLAMP, degausser coil mtg (16 & E16TS-929)
	42D65572A17	CLAMP, degausser coil mtg (18 & F18TS-929)
	15S10183A15	CONNECTOR, plug: 1 contact; less contact (AFT - E16 & F18TS-929)
P-7	15S10183A14	CONNECTOR, plug: 10 contact; less contacts (pre set - E16, F18TS-929)
	15S10183A16	CONNECTOR, recept: 1 contact; less contact (AFT - E16 & F18TS-929)
S-7	15S10183A13	CONNECTOR, recept: 10 contact; less contacts (pre set - E16, F18TS-929)
	39S10184A02	CONTACT, plug: for AFT & P-7 connector plug
	39S10184A01	CONTACT, recept: for AFT & S7 connector recept
	14B65953A01	INSULATOR, armite: CRT wire retainer (16 &E16TS-929)
	14B70029A02	INSULATOR, fibre: vol ctrl mtg (R301)
	2S10054A36	NUT, spring: clip on; CRT wire retainer (16 &E16TS-929)
	42C67092A06	RETAINER, wire: CRT; less screw & clip on nut (16 &E16TS-929)
	3S136803	SCREW, tpg: 8-18 x 2-3/4; CRT wire retainer (16 &E16TS-929)
	3S135474	SCREW, mch: 1/4-20 x 7/16; CRT mtg (16 & E16TS-929)
	3B68012A03	SCREW, mch: 1/4-20 x 5/8; CRT mtg (18 & F18TS-929)
	26B68206A01	SHIELDED, pilot light: UHF & VHF (E16 & F18TS-929)
	26B65584B05	SHIELD, pilot light: pre set (E16TS-929)
	26B70509A01	SHIELD, pilot light: pre set (F18TS-929)
	9V68635A54	SOCKET, pilot light: UHF/VHF; incls lead & contact (E16 & F18TS-929)

REF.	PART NO.	DESCRIPTION
	9V68645A88	SOCKET, pilot light: pre set; incls disconnect (E16TS-929)
	9B70488A01	SOCKET, pilot light: pre set; incls lead (F18TS-929)
	41A70031A02	SPRING, CRT shield mtg (16 & E16TS-929)
	41A70031A01	SPRING, CRT shield mtg (18 & F18TS-929)
	42B70163A01	STRAP, CRT mtg; 4 used (16 & E16TS-929)
	4S7657	WASHER, lock: CRT wire retainer (16 & E16TS-929)

CABINET PARTS FOR WP457GN, WP464GW & WP467GWA

REF.	PART NO.	DESCRIPTION
	85D67625A06	ANTENNA, UHF: "bow tie"
	16P65175A46	CABINET FRONT ASSEMBLY, incls overlays (WP457GN)
	16P65175A47	CABINET FRONT ASSEMBLY, incls overlays (WP464GW)
	16P65175A48	CABINET FRONT ASSEMBLY, incls overlays, dial light windows, instamatic color lens (WP467GWA)
	16B69906A12	CART, TV: complete (WP464GW)
	16B69906A10	CART, TV: complete (WP467GWA)
	42S10122A01	CLIP, secures line cord to cab back
	30S183A01	CORD, AC line
	15P65175A44	COVER, cabinet back: incls shields; less all other components (WP457GN)
	15P65175A45	COVER, cabinet back: incls shields; less all other components (WP464GW & WP467GWA)
	55D70024A02	HANDLE
	14D70028B01	INSULATOR, nylon: handle mtg
	36D70167A04	KNOB, channel selector: UHF
	36D70167A03	KNOB, channel selector: VHF
	36D70012A01	KNOB, control: brightness-contrast-vertical
	36D68493A30	KNOB, control: on-off vol (WP457GN)
	36D68493A26	KNOB, control: on-off vol (WP464GW & WP467GWA)
	36D70166A02	KNOB, fine tune: UHF
	36D70166A01	KNOB, fine tune: VHF
	36D70045B01	KNOB, slide: hue & intensity
	61B70490A01	LENS, instamatic color (WP467GWA)
	29K730154	LUG, spade: antenna
	2S115123	NUT, hex: 10-32; handle mtg
	2S10101A23	NUT, push-on: dial light window (WP467GWA)
	13C70014A04	OVERLAY, control: "on-vol" (WP457GN)
	13C70014A01	OVERLAY, control: "on-vol" (WP464GW)
	13C70014A02	OVERLAY, control: "instamatic color & on vol" (WP467GWA)
	13C70154A01	OVERLAY, control panel: UHF/VHF (WP457GN & WP464GW)
	13C70154A02	OVERLAY, control panel: UHF/VHF AFT (WP467GWA)
	13C70013A02	OVERLAY, nameplate & slide control "Quasar - hue & intensity"
	36D70491A03	PUSHBUTTON, instamatic color (pre set - WP467GWA)
	3S138107	SCREW, tpg: 8-18 x 11/16; cab back mtg
	3S135222	SCREW, tpg: 8-18 x 3/4; interlock
	50D68384A02	SPEAKER, 4" PM 16Ω VC
	85D70023A02	TUBE ROD & BALL ASSEMBLY: VHF ant; mtg brkt & slvg
	4S7582	WASHER, flat: metal; handle mtg
	61D70147A01	WINDOW, dial light: UHF/VHF (WP467GWA)

CABINET PARTS FOR WP553GN, WP554GW & WP563GWA

REF.	PART NO.	DESCRIPTION
	85D67625A06	ANTENNA, UHF: "bow tie"
	16P56175A51	CABINET FRONT ASSEMBLY, incls overlays (WP553GN)
	16P65175A52	CABINET FRONT ASSEMBLY, incls overlays (WP554GW)

REF.	PART NO.	DESCRIPTION	REF.	PART NO.	DESCRIPTION
	16P65175A53	CABINET FRONT ASSEMBLY, incls overlays, dial light windows & color instamatic lens (WP563GWA)		36C70056A01	KNOB, slide: hue & intensity
	16B69906A13	CART, TV: complete (WP554GW)		61B70490A01	LENS, instamatic color (pre-set) (WP563GWA)
	16B69906A11	CART, TV: complete (WP563GWA)		29K730154	LUG, spade: antenna
	42S10152A02	CLIP, push-on: speaker mtg		2S115123	NUT, hex: 10-32; handle mtg
	42S10122A01	CLIP, secures line cord to cab back		2S10101A23	NUT, push-on: dial light window (WP563GWA)
	30S183A01	CORD, AC line		13C70289A01	OVERLAY, control: "bright-contrast-vertical" (WP553GN & WP554GW)
	15P65175A49	COVER, cabinet back: incls shield less all other components (WP553GN)			
	15P65175A50	COVER, cabinet back: incls shields; less all other components (WP554GW & WP563GWA)		13C70289A02	OVERLAY, control: "instamatic color-bright-contrast-vertical" (WP563GWA)
	55D70024A01	HANDLE		13C70058A03	OVERLAY, control: "on-vol" (WP553GW)
	14D70028B01	INSULATOR, nylon: handle mtg		13C70058A01	OVERLAY, control: "on-vol"(WP554GW)
	36D70047A02	KNOB, channel selector: UHF; incls dial scale (WP553GN & WP554GW)		13D70072A01	OVERLAY, control panel: "UHF/VHF vol AFT" (WP563GWA)
	36D70047A06	KNOB, channel selector: UHF; incls dial scale (WP563GWA)		13C70075A05	OVERLAY, nameplate & slide control: "Quasar-intensity-hue" (WP553GN & WP554GW)
	36D70047A01	KNOB, channel selector: VHF; incls dial scale (WP553GN & WP554GW)		13C70075A04	OVERLAY, nameplate & slide control "Quasar-intensity-hue" (WP563GWA)
	36D70047A05	KNOB, channel selector: VHF; incls dial scale (WP563GWA)		36D70491A05	PUSHBUTTON, instamatic color (pre-set - WP563GWA)
	36D70057A01	KNOB, control: brightness-contrast-vertical		3S138007	SCREW, tpg: 8-18 x 11/16; cab back mtg
	36D70048A02	KNOB, fine tune: UHF (WP553GN & WP554GW)		3S135222	SCREW, tpg: 8-18 x 3/4; interlock
	36D70048A04	KNOB, fine tune: UHF (WP563GWA)		50D69062A04	SPEAKER, 3 x 5 PM 16Ω VC
	36D70048A01	KNOB, fine tune: VHF (WP553GN & WP554GW)		85D70023A01	TUBE ROD & BALL ASSEMBLY, VHF ant: incls mtg brkt & slvg
	36D70048A03	KNOB, fine tune: VHF (WP563GWA)		4S7582	WASHER, flat: metal; handle mtg
	36D68493A25	KNOB, on-off volume		61B70030A01	WINDOW, dial light: UHF/VHF (WP563GWA)

CTV7 & 8

REF.	PART NO.	DESCRIPTION	REF.	PART NO.	DESCRIPTION
UHF WHEEL AND SWITCH PLATE ASSEMBLY REPLACEMENT PARTS (ATS938)				43K471633	BEARING, ball
17	61D70420A17	CARD, channel identification: numbers 2 thru 13 & incls 1 with letter "U"		1A41270C01	CLUTCH & DISC ASSEMBLY: incls mtg screw
18	61D70420A16	CARD, channel identification: numbers 14 thru 83 & incls 8 blank		33D70741A01	DECAL, UHF chan numbers: for pushbutton; #14 thru 83 incls 8 blank
1	45X97023A06	CAM, trigger		44B42707A01	GEAR, clutch face
14	45X97023A08	CONTACT, for slider carrier		1B42770A04	GEAR & SHAFT ASSEMBLY: incls worm & dog gear
2	49X97032A15	UHF WHEEL, complete: less mtg plate		2S121771	NUT, hex: 3/8-32; manual tun shaft mtg
3	45X97022A19	LEVER, control: UHF/VHF		38C41275C01	PUSHBUTTON
4	64X97021A12	PLATE, mtg: complete; incls ctrl lever, switches & wiper assembly		1V68650A01	SCALE, dial: UHF; manual tuning: incls drum
5	64X97021A13	PLATE, mtg: incls staked on ctrl lever shaft only		3S138468	SCREW, mch: 6-32 x 5/16; UHF dial scale assembly mtg
6	64X97021A11	POTENTIOMETER PLATE		1V562830	SCREW & NUT ASSEMBLY: adjustable
7	39C70513A01	SLIDER SCREW RETAINER AND POT CONTACT		47A42771A04	SHAFT, manual tuning: incls gear & mtg bushing
8	42X97028A06	RING, collar	**UHF TUNER TT-656**		
9	3X97030A31	SCREW, pot plate mtg			
10	3A70401A01	SCREW, set (USE 3-40679B01)		- - - -	NOT SUPPLIED SEPARATELY. PART OF PANEL "A"
11	3X97030A30	SCREW, slider carrier: less slider & contact			
12	3X97030A32	SCREW, switch & wiper assembly mtg	**UHF TUNER TT-657**		
13	45X97023A07	SLIDER CARRIER, less contact	**ELECTRICAL PARTS**		
19	40X97002A14	SWITCH, leaf: UHF/VHF (SW1 & SW3)	D-1	48P65112A73	DIODE: mixer
20	40X97002A13	SWITCH, leaf: UHF/VHF (SW2 & SW4)	D-2	48X90233A09	DIODE: AFT; 1S1923A
15	4X97039A21	WASHER, "C": retains drum to mtg plate	Q-1	48S134902	TRANSISTOR: A1E
16	4X97039A20	WASHER, "C": ctrl lever mtg (UHF/VHF)	**MECHANICAL PARTS**		
21	39X97007A06	WIPER ASSEMBLY, complete		77P65175A88	UHF TUNER: TT657; complete: incls plug assembly less spur gear
UHF PUSHBUTTON DRIVE ASSEMBLY (TS-938)				44B40680B01	GEAR, spur: incls set screw
	77U42901B84	DRIVE & PUSHBUTTON ASSEMBLY: complete; incls pushbutton, gear clutch & bushing assembly, tuning shaft; less UHF tuner, & dial scale drum assembly	**UHF TUNER TT-658**		
			ELECTRICAL PARTS		
			D-1	48P65112A73	DIODE: mixer

REF.	PART NO.	DESCRIPTION	REF.	PART NO.	DESCRIPTION
D-2	48P65148A02	DIODE: AFC	2	44X90277A02	GEAR, fine tune: incls cam assembly
Q-1	48P65174A24	TRANSISTOR: SPS4145	3	3X90262A04	SCREW, gear head: fine tune

MECHANICAL PARTS

REF.	PART NO.	DESCRIPTION	REF.	PART NO.	DESCRIPTION
	77P65175A88	UHF TUNER: TT658; complete; incls plug assembly less spur gear	4	41X90271A07	SPRING, channel selector shaft ground
	44B40680B01	GEAR, spur: incls set screw	5	4X90259A05	WASHER, "C": secures fine tune gear to chan sel shaft

VHF TUNER OPTT-429

ELECTRICAL PARTS

VHF TUNER OPTT-437

ELECTRICAL PARTS

CAPACITORS

MISCELLANEOUS ELECTRICAL PARTS

REF.	PART NO.	DESCRIPTION	REF.	PART NO.	DESCRIPTION
D-1	48P65173A77	DIODE, AFT: BA142-01	C-5	21X90303A33	8.2 pf 5% N470
E-1	24P65173A80	ANTENNA INPUT FILTER: complete	C-7	21P65173A75	47 pf 5% feed-thru
SW-6	40P65173A79	SWITCH, UHF B+ and pilot light	C-8	21X90303A34	15 pf 5% N750
			C-9	21X90303A35	68 pf 5% N470
COILS & CHOKES			C-10	21X90303A35	68 pf 5% N470
			C-11	21X90303A36	1000 pf +100-20%
L-22	24P65148A05	COIL, master oscillator: incls core	C-12	21X90303A37	11 pf 5% NPO
L-23	24P65173A82	COIL, mixer: incls core			

MISCELLANEOUS ELECTRICAL PARTS

REF.	PART NO.	DESCRIPTION	REF.	PART NO.	DESCRIPTION
TRANSISTORS			E-1	24P65173A80	ANTENNA INPUT ASSEMBLY
			IC-1	51X90305A01	MODULE: P020B
Q-1	48P65146A61	RF AMP: SE5020	SW-6	40P65173A79	SWITCH, UHF B+ & pilot light
Q-2	48P65173A78	MIXER: SE5030			
Q-3	48P65146A63	OSCILLATOR: SE1010	COILS		

REF.	PART NO.	DESCRIPTION
L-22	24X90243A49	OSCILLATOR: incls core
L-23	24X90243A50	MIXER: incls core

MECHANICAL PARTS

REF.	PART NO.	DESCRIPTION	REF.	PART NO.	DESCRIPTION
	77V68649A46	VHF TUNER, complete; incls plugs convertor, cable, and all other outboard component parts less mtg brkt		77V68649A46	VHF TUNER, OPTT437: complete; incls plugs, convertor cable & all other outboard component parts; less mtg brkt
1	1P65120A46	DRIVE LINK ASSEMBLY, incls 10 & 14 tooth gears	1	45X90304A01	DRIVE LINK ASSEMBLY: incls 10 & 14 tooth gears
2	47P65146A44	GEAR, fine tune: incls bushing, clutch & metal gear	2	44X90277A07	GEAR, fine tune: incls bushing, clutch & metal gear
3	44P65146A15	GEAR, fine tune: 14 tooth; engages gear head screws	3	44X90277A05	GEAR, 10 tooth: idler; drive link
4	44P65146A16	GEAR, idler: 10 tooth; drive link	4	44X90277A06	GEAR, 14 tooth: drive link; engages gear head screws
5	3P65145A82	SCREW, gear head: fine tune	5	3X90262A07	SCREW, gear head: fine tune
6	47P65173A81	SHAFT, channel selector: incls detent wheel & pre-set screw & coil assembly	6	47X90276A21	SHAFT, channel selector: incls detent wheel, pre-set screw, coil & osc stator assembly
7	42P65120A43	SPRING, bias: retains shaft to front plate	7	41X90271A10	SPRING, bias: retains channel selector shaft to front plate
8	41A747735	SPRING, chan sel shaft grounding	8	41X90271A11	SPRING, channel selector shaft ground
9	42P65145A84	SPRING, detent: chan sel shaft (located on front plate)	9	41X90271A18	SPRING, channel selector shaft ground (SW6)
10	41P65120A42	SPRING, drive link return	10	41X90271A12	SPRING, detent: channel selector shaft (on front plate)
11	4P65114A53	WASHER, "C" (secures fine tune gear to chan sel shaft)	11	41X90271A13	SPRING, drive link return
			12	4X90259A07	WASHER, "C": fine tune gear

MISCELLANEOUS ELECTRICAL PARTS

VHF TUNER AOPTT-439

ELECTRICAL PARTS

COILS & CHOKES

REF.	PART NO.	DESCRIPTION	REF.	PART NO.	DESCRIPTION
D-1	48X90233A08	DIODE, AFT: S2085	L-22	24X90243A43	MASTER OSCILLATOR, incls core
E-1	24P65171A94	ANTENNA INPUT ASSEMBLY	L-25	24P65173A02	MIXER, incls core
E-2	76X90301A01	FERRITE BEAD			
SW-1C	59X90285A03	OSCILLATOR STATOR ASSEMBLY	TRANSISTORS		
SW-6	40P65149A66	SWITCH, UHF B+			
			Q-1	48X90232A21	RF AMP, SE5020
COILS & CHOKES			Q-2	48X90232A24	MIXER: SE5031
			Q-3	48X90232A22	OSCILLATOR, SE1010
L-14	24X90243A32	COIL, UHF INPUT, shunt			
L-15	24X90243A33	COIL, UHF INPUT, series	SWITCHES		
L-24	24P65149A41	COIL, mixer: incls core			
L-40	24X90243A34	COIL, FM trap: incls core & C6	SW-6	40P65174A23	SWITCH, program: (channel stop skip)
L-45	24X90243A31	CHOKE, RF			

TRANSISTORS

REF.	PART NO.	DESCRIPTION
Q-1	48S137158	RF: A6E
Q-2	48S134950	MIXER: A2H
Q-3	48S134949	OSCILLATOR: A2G

REF.	PART NO.	DESCRIPTION
	77V68650A89	VHF TUNER: complete; incls 6 contact connector plug, convertor cable, and all other outboard component parts: less mtg brkt

MECHANICAL PARTS

REF.	PART NO.	DESCRIPTION
	77V68649A46	VHF TUNER, complete; incls plugs, convertor cable, all all other outboard component parts, less mtg brkt
1	47X90276A05	FRONT PLATE ASSEMBLY, complete: incls chan sel shaft fine tune, idler & drive gears & pre-set screw & coil assembly (SW1A & B)

MECHANICAL PARTS

REF.	PART NO.	DESCRIPTION
1	45X90304A01	DRIVE LINK ASSEMBLY, incls 10 & 14 tooth gears

REF.	PART NO.	DESCRIPTION	REF.	PART NO.	DESCRIPTION
2	44X90277A07	GEAR, fine tune: incls bushing, clutch & metal gear		41C70756A02	HINGE, spring: ctrl door; incls mtg pin
3	44X90277A05	GEAR, 10 tooth: idler; drive link		15C69376B02	HOLDER, remote transmitter
4	44X90277A06	GEAR, 14 tooth: drive link; engages gear head screw		36C70318A02	KNOB, channel selector: UHF/VHF
				36D70338B04	KNOB, control: AFT
5	3X90262A07	SCREW, gear head: fine tune		36D69941A11	KNOB, control: on-off/volume
6	47X90276A25	SHAFT, channel selector: incls detent wheel, pre-set screw, coil & osc stator assembly (SW1)		36C70738A01	KNOB, control: vertical
				36C70311A02	KNOB, fine tune: UHF/VHF
				36D69981A03	KNOB, slide: intensity & hue
7	41X90271A10	SPRING, bias: retains channel selector shaft to front plate		61D70729A01	LENS, indicator: "push on vol/ insta-matic/power tune"
8	41X90271A11	SPRING, channel selector shaft grounding		2S7981	NUT, push on: indicator lens
9	41X90271A18	SPRING, channel selector shaft grounding		13C70775A01	OVERLAY, control door: "Quasar-Remote"
10	41X90271A12	SPRING, detent: channel selector shaft (on front plate)		13C70726A01	OVERLAY, indicator lens
				36D70731A02	PUSHBUTTON: insta-matic
11	41X90271A13	SPRING, drive link return		36C70383A02	PUSHBUTTON: power tune
12	4X90259A07	WASHER, "C" fine tune gear		3D66303A18	SCREW, tpg: 8-15 x 3/4; CRT bezel
		CABINET PARTS		3D66303A19	SCREW, tpg: 8-15 x 13/16; interlock
	MODEL WT685HW & WT685HWC			3S138469	SCREW, tpg: 8-18 x 3/4 blk ox; esc ctrl panel
	85D67625A08	ANTENNA, UHF: "bow tie"		50D68384A02	SPEAKER, 4" PM 16Ω VC
	85D68375A03	ANTENNA, VHF: "dipole"		41B65987B09	SPRING, CRT bezel grnd
	13E70323A03	BEZEL, CRT		41C70069A03	SPRING, speaker grille mtg
	75B67953A02	BUMPER, cab foot			
	16E70084A11	CABINET, TV: metal; walnut: incls slide assem		**MODEL WT688HW**	
	42S10122A01	CLIP, secures line cord to cab back		13E70323A04	BEZEL, CRT
				16E70662A01	CABINET, TV: wood, walnut
	42S10152A02	CLIP, push-on: spkr mtg		42S10122A01	CLIP, secures line cord to cab back
	42C67787A04	CLIP, spring: cab back mtg		42S10152A02	CLIP, push-on: spkr mtg
	42B70093A02	CLIP, spring hinge: ctrl door		42C67787A04	CLIP, spring: cab back mtg
	15E70132B04	COVER, cabinet back: less all other components		42B70093A02	CLIP, spring hings: ctrl door
	15E68244A02	COVER, CRT: rear (USE 15D67156A02)		15V68650A10	COVER, cabinet back: complete; incls line cord, ant. board cover & rear CRT cover
	30S183A20	CORD, AC line		15E68244A02	COVER, CRT: rear (USE 15D67156A02)
	15P65175A79	DOOR, control: incls overlay		30S183A20	CORD, AC line
	13P65175A78	ESCUTCHEON, control panel: incls overlays, dial windows; less ctrl door & spkr grille		15P65175A81	DOOR, control: incls overlay
				13P65175A80	ESCUTCHEON, control panel; incls overlays, dial windows, less ctrl door & spkr grille
	13E70733A01	GRILLE, speaker		1V68650A09	GRILLE, speaker: incls cloth panel overlay
	36D69941A08	KNOB, on-off volume		36D69941A12	KNOB, on-off volume
	36C70600A01	KNOB, UHF fine tune		36C70600A01	KNOB, UHF fine tune
	36C70738A01	KNOB, vertical		36C70738A01	KNOB, vertical
	36D70324A08	KNOB, VHF channel selector		36D70324A11	KNOB, VHF channel selector
	36D70324A07	KNOB, VHF fine tune		36D70324A09	KNOB, VHF fine tune
	13C69992A06	OVERLAY, control: "video peak, tone, hue, etc"		13C69992A06	OVERLAY, control: "video peak, tone, hue, etc"
	13D70740A01	OVERLAY, control panel: translucent; "on-vol/VHF FT/ INSTA-MATIC, etc"		13D70740A02	OVERLAY, control panel: translucent; "on-vol/VHF FT/INSTA-MATIC, etc"
	13B70774A01	OVERLAY, control door: "Quasar"		13B70774A01	OVERLAY, control door: "Quasar"
	36D70731A01	PUSHBUTTON: INSTA-MATIC & AFT		36D70731A03	PUSHBUTTON: INSTA-MATIC & AFT
	34D69996A07	SCALE, dial: VHF		34D69996A07	SCALE, dial: VHF
	3D66303A19	SCREW, tpg: 8-15 x 13/16; interlock		3D66303A19	SCREW, tpg: 8-15 x 13/16; interlock
	3S138469	SCREW, tpg: 8-18 x 3/4 phl flt blk ox; ctrl esc mtg		3S138469	SCREW, tpg: 8-18 x 3/4 phl flt blk ox; ctrl esc mtg
	50D68384A02	SPEAKER 4" PM 16Ω VC		1V68650A07	SLIDE ASSEMBLY: chassis; complete; incls slide lock spring
	41B67987B09	SPRING, CRT bezel grnd		50D68384A02	SPEAKER 4" PM 16Ω VC
	41B70094A01	SPRING, chassis slide lock		41C70069A03	SPRING, speaker grille mtg
	41C70069A03	SPRING, speaker grille mtg		61A70474A02	WINDOW, dial: UHF/VHF
	61A70474A02	WINDOW, dial: UHF & VHF		**MODEL WU904HW, HWC, WU906HW, HWC, WU907HS & WU908HP, HPC**	
	MODEL TT687HW			13E70276A01	BEZEL, CRT: (WU904HW,HWC WU906HW, HWC & WU908HP,HPC)
	85D67625A08	ANTENNA, UHF: "bow tie"		13E70276A05	BEZEL, CRT: (WU907HS)
	85D68375A03	ANTENNA, VHF: "dipole"		16E70514A02	CABINET, TV: console; walnut (WU904HW)
	13E70323A03	BEZEL, CRT		16E70514A04	CABINET, TV: console; walnut (WU904HWC)
	75B67953A02	BUMPER, cab foot		16E70690A01	CABINET, TV: console; walnut (WU906HW)
	16E70084A11	CABINET, TV: metal; walnut: incls slide assembly		16E70690A02	CABINET, TV: console; walnut (WU906HWC)
	42S10122A01	CLIP, secures line cord to cab back		16E70719A01	CABINET, TV: console; maple (WU907HS)
	42S10152A02	CLIP, push-on: spkr mtg		16E70515A03	CABINET, TV: console; pecan (WU908HP, HPC)
	42C67787A04	CLIP, spring: cab back mtg		55C63659A23	CASTER: less grip neck socket (WU904HW, HWC, WU907HS, UW908HP, HPC)
	15E70132B04	COVER, cabinet back: less all other components		42S10122A01	CLIP, secures line cord to cab back
	15D68244A02	COVER, CRT: rear (USE 15D67156A02)			
	30S183A20	CORD, AC line			
	15P65175A94	DOOR, control: incls window, pins, overlay, & hinge springs			
	13P65176A32	ESCUTCHEON, control panel: incls ctrl indicator lens, door inserts & indicator overlay: less ctrl door & spkr grille			
	13D70733A01	GRILLE, speaker: less mtg spring			

REF.	PART NO.	DESCRIPTION
	42S10152A02	CLIP, push-on: spkr mtg
	42B70245A01	CLIP, spring catch: CRT bezel mtg
	42C67787A04	CLIP, spring: cab back mtg
	42B70093A02	CLIP, spring hinge: ctrl door
	15V68669A03	COVER, cabinet back; complete; incls line cord, CRT cover & ant board cover
	15E68244A05	COVER, CRT: rear
	30S183A20	CORD, AC line
	15P65175A84	DOOR,control:incls overlay(WU904HW, HWC, WU906HW, HWC, WU908HP,HPC)
	15P65175A85	DOOR, control: incls overlay (WU907HS)
	13P65175A83	ESCUTCHEON, control panel: incls overlays, dial windows; less ctrl door & spkr grille
	13E70734A01	GRILLE, speaker (WU904HW, HWC, WU906HW, HWC, WU908HP, HPC)
	13V68669A08	GRILLE, speaker: incls cloth panel overlay (WU907HS)
	55C62243B07	KEY, esc: decorative (WU907HS)
	36D69941A08	KNOB, on-off volume
	36C70600A01	KNOB, UHF fine tune
	36C70738A01	KNOB, vertical
	36D70324A08	KNOB, VHF channel selector
	36D70324A07	KNOB, VHF fine tune
	16B70115A01	LEG, cabinet (WU904HW)
	16B70115A02	LEG, cabinet (WU904HWC)
	13C69992A06	OVERLAY, control: "video peak, tone. hue, etc)
	13D70740A01	OVERLAY, control panel: translucent; "on-vol/VHF FT/INSTA-MATIC, etc"
	13B70774A03	OVERLAY,control door:"Quasar"(WU904HW HWC,WU906HW,HWC, WU908HP,HPC)
	13B70774A04	OVERLAY, control door: "Quasar" (WU907HS)
	16D70743A03	OVERLAY, decorative panel: plastic; cab front (WU908HP, HPC)
	36D70731A01	PUSHBUTTON: INSTA-MATIC & AFT
	64A67158A02	PLATE, metal: leg mtg (WU904HW, HWC)
	55C62243B08	PULL, door: decorative (WU906HW, HWC)
	55C62243B06	PULL, door: decorative (WU907HS)
	34D69996A07	SCALE, dial: VHF
	3D66303A19	SCREW, tpg: 8-15 x 13/16; interlock
	3S138469	SCREW, tpg: 8-18 x 3/4 phl flt blk ox; ctrl esc mtg
	1V68643A47	SLIDE ASSEMBLY: chassis; complete: incls slide lock spring
	55C63659A24	SOCKET, grip neck: for caster (WU904HW,HWC,WU906HW,HWC,WU908HP,HPC)
	50D67337A01	SPEAKER, 4 x 6 PM 16Ω VC
	41C70069A03	SPRING, spkr grille mtg
	16C70689A01	STRETCHER, cab leg (WU904HW)
	16C70689A02	STRETCHER, cab leg (WU904HWC)
	61A70474A02	WINDOW, dial UHF & VHF

MODEL WT933HW, HWC

REF.	PART NO.	DESCRIPTION
	85D67625A08	ANTENNA, UHF: "bow tie"
	85D68375A03	ANTENNA, VHF: "dipole"
	13E70276A04	BEZEL, CRT
	75B67953A02	BUMPER, cab foot
	16E70462A01	CABINET, TV: metal; walnut incls slide assembly
	42S10122A01	CLIP, secures line cord to cab back
	42S10152A02	CLIP, push on: spkr mtg
	42B70245A01	CLIP, spring catch: CRT bezel mtg
	42C67787A04	CLIP, spring: cab back mtg
	42B70093A02	CLIP, spring hinge: ctrl door
	15P65175A82	COVER, cabinet back: incls ant. board cover, less all other components
	15E68244A05	COVER, CRT: rear
	30S183A20	CORD, AC line
	15P65175A84	DOOR, control: incls overlay
	13P65175A83	ESCUTCHEON, control panel: incls overlays, dial windows; less ctrl door & spkr grille
	13E70734A01	GRILLE, speaker
	36D69941A08	KNOB, on-off volume
	36C70600A01	KNOB, UHF fine tune
	36C70738A01	KNOB, vertical
	36D70324A08	KNOB, VHF channel selector
	36D70324A07	KNOB, VHF fine tune
	16B69964A01	LEG, cabinet
	13C69992A06	OVERLAY, control: "video peak, tone, hue, etc"

REF.	PART NO.	DESCRIPTION
	13D70740A01	OVERLAY, control panel: translucent; "on-vol/VHF FT/INSTA-MATIC, etc"
	13B70774A03	OVERLAY, control door: "Quasar"
	36D70731A01	PUSHBUTTON: INSTA-MATIC & AFT
	34D69996A07	SCALE, dial: VHF
	3D66303A19	SCREW, tpg: 8-15 x 13/16; interlock
	3S138469	SCREW, tpg: 8-17 x 3/4 phl flt blk ox; ctrl esc mtg
	50D67337A01	SPEAKER, 4 x 6 PM 16Ω VC
	41B70094A01	SPRING, chassis slide lock
	41C70069A03	SPRING, spkr grille mtg
	61A70474A02	WINDOW, dial: UHF & VHF

MODEL TT934HW

REF.	PART NO.	DESCRIPTION
	85D67625A08	ANTENNA, UHF: "bow tie"
	85D68375A03	ANTENNA, VHF: "dipole"
	13E70276A04	BEZEL, CRT
	75B67953A02	BUMPER, cab foot
	16E70462A01	CABINET, TV: metal; walnut: incls slide assembly
	42S10122A01	CLIP, secures line cord to cab back
	42S10152A02	CLIP, push on: spkr mtg
	42B70245A01	CLIP, spring catch: CRT bezel mtg
	42C67787A04	CLIP, spring: cab back mtg
	15P65175A99	COVER, cabinet back: incls ant board cover & remote transmitter hsg: less all other components
	15D68244A05	COVER, CRT: rear
	30S183A20	CORD, AC line
	15P65175A97	DOOR, control: incls window, pins, overlay, & hinge springs
	13P65175A96	ESCUTCHEON, control panel: incls ctrl indicator lens, door inserts & indicator overlay: less ctrl door & spkr grille
	13E70734A01	GRILLE, speaker: less mtg spring
	41P65175A95	HINGE, spring: ctrl door; incls mtg pin
	15C69376B02	HOLDER, remote transmitter
	36C70318A02	KNOB, channel selector: UHF/VHF
	36D70338B04	KNOB, control: AFT
	36D69941A11	KNOB, control: on-off/volume
	36C70738A01	KNOB, control: vertical
	36C70311A02	KNOB, fine tune: UHF/VHF
	36D69981A03	KNOB, slide: intensity & hue
	16B69964A01	LEG, cabinet
	61D70729A01	LENS, indicator: "push on vol/insta-matic/power tune"
	2S7981	NUT, push on: indicator lens
	13C70775A02	OVERLAY, control door: "Quasar-remote"
	13C70726A01	OVERLAY, indicator lens
	36D70731A02	PUSHBUTTON: insta-matic
	36D70383A02	PUSHBUTTON: power tune
	3D66303A18	SCREW, tpg: 8-15 x 3/4; CRT bezel
	3D66303A19	SCREW, tpg: 8-15 x 13/16; interlock
	3S138469	SCREW, tpg: 8-18 x 3/4 blk ox; esc ctrl panel
	50D67337A01	SPEAKER, 4 x 6 PM 16Ω VC
	41C70069A03	SPRING, speaker grille mtg

MODEL TU905HW, TU944HW, TU945HS, TU946HK

REF.	PART NO.	DESCRIPTION
	13E70276A01	BEZEL, CRT (TU905HW, TU944HW, TU946HK)
	13E70276A05	BEZEL, CRT (TU945HS)
	16E70514A02	CABINET, TV: console; walnut (TU905HW)
	16E70693A01	CABINET, TV: console; walnut (TU944HW)
	16E70667A01	CABINET, TV: console; maple (TU945HS)
	16E70692A01	CABINET, TV: console; oak (TU946HK)
	55C63659A23	CASTER: less grip neck socket (TU944HS, TU945HS, TU946HK)
	42S10122A01	CLIP, secures line cord to cab back
	42S10152A02	CLIP, push on: spkr mtg
	42B70245A01	CLIP, spring catch: CRT bezel mtg

REF.	PART NO.	DESCRIPTION
	15V68669A59	COVER, cabinet back: complete; incls ant board cover, remote trans hsg, CRT rear cover & line cord (TU905HW)
	15V68669A65	COVER, cabinet back: complete; incls ant board cover, remote trans hsg, CRT rear cover & line cord (TU944HW, TU945HS, TU946HK)
	15D68244A05	COVER, CRT: rear
	30S183A20	CORD, AC line
	15P65175A97	DOOR, control: incls window, pins, overlay & hinge springs (TU905HW, TU944HW, TU946HK)
	15P65175A98	DOOR, control: incls window, pins, overlay & hinge springs (TU945HS)
	13P65175A96	ESCUTCHEON, control panel: incls ctrl indicator lens, door inserts & indicator overlay: less ctrl door & spkr grille
	13E70734A01	GRILLE, speaker: less mtg spring (TU905HW, TU944HW, TU946HK)
	13V68669A68	GRILLE, speaker: incls cloth panel overlay: less mtg spring (TU945HS)
	41C70756A02	HINGE, spring: ctrl door; incls mtg pin
	15C69376B02	HOLDER, remote transmitter
	36C70318A02	KNOB, channel selector: UHF/VHF
	36D70338B04	KNOB, control: AFT
	36D69941A11	KNOB, control: on-off/volume
	36C70738A01	KNOB, control: vertical
	36C70311A02	KNOB, fine tune: UHF/VHF
	36D69981A03	KNOB, slide: intensity & hue
	16B70115A01	LEG, cabinet: (TU905HW)
	61D70729A01	LENS, indicator: "push on vol/ insta-matic/power tune"
	2S7981	NUT, push on: indicator lens
	13C70775A02	OVERLAY, control door: "Quasar-remote" (TU905HW, TU944HW, TU946HK)
	13C70775A03	OVERLAY, control door: "Quasar-remote" (TU945HS)
	13C70726A01	OVERLAY, indicator lens
	13D69957A36	OVERLAY, cloth panel: spkr panel
	64A67158A02	PLATE, metal: leg mtg (TU905HW)
	36D70731A02	PUSHBUTTON: insta-matic
	36D70383A02	PUSHBUTTON: power tune
	3D66303A19	SCREW, tpg: 8-15 x 13/16; interlock
	3S138469	SCREW, tpg: 8-18 x 3/4 blk ox; esc ctrl panel
	1V68643A47	SLIDE ASSEMBLY: chassis; complete: incls slide lock spring
	55C63659A24	SOCKET, grip neck: for caster
	50D67337A01	SPEAKER, 4 x 6 PM 16Ω VC
	41C70069A03	SPRING, spkr grille mtg
	16E70689A01	STRETCHER, cab leg: (TU905HW)

UHF TUNER & CONTROL PANEL "A"

REF.	PART NO.	DESCRIPTION
	1Y68663A99	UHF TUNER CONTROL PANEL "A": complete

ELECTRICAL PARTS

REF.	PART NO.	DESCRIPTION
	- - - -	UHF TUNER: TT651; not replaceable (supplied only as part of panel "A": complete)
D-5A	48S137272	DIODE, Zener: D4J
D-8A	48S137017	DIODE, Zener: D1T

TRANSISTORS

REF.	PART NO.	DESCRIPTION
Q-4A	48S137300	AGC AMP: A8B
Q-5A	48S134944	AFC AMP: W1A
Q-6A	48S134932	IF AMP: A1V
Q-7A	48S137127	AGC DRIVER: P2S

TRANSFORMERS

REF.	PART NO.	DESCRIPTION
T-2A	24D68501A17	IF OUTPUT

MECHANICAL PARTS

REF.	PART NO.	DESCRIPTION
	29S10134A29	CONNECTOR, recept: module panel mtg
	15S10183A15	CONNECTOR, plug: 1 contact; less contact AFT

REF.	PART NO.	DESCRIPTION
	39S10184A02	CONTACT, plug: for connector 15S10183A15)
	9C67349A07	RECEPT, phono: UHF input (on tuner brkt)
	9C67349B02	RECEPT, phono: IF output (on module panel)

IF AUDIO PANEL "BA"

REF.	PART NO.	DESCRIPTION
	1Y68662A57	IF AUDIO PANEL "BA": complete

ELECTRICAL PARTS

COILS & CHOKES

REF.	PART NO.	DESCRIPTION
L-1	24D68801A02	COIL, compensating: 450 uh
L-2	24D66772A12	CHOKE, resonant: 7.5 uh
L-3	24D68517A20	COIL, quadrature
L-4	24D68501A19	COIL, convertor secondary
L-5	24D69707A13	COIL, hi pass filter: .326 uh
L-6	24D68501A19	COIL, 39.75 MHz trap
L-7	24D69707A13	COIL, hi pass filter: .326 uh
L-8	24D68501A18	COIL, 47.25 MHz trap
L-9	24D68501A16	COIL, 41.25 MHz trap
L-11	24D66772A12	CHOKE, resonant: 7.5 uh
L-12	24D68588A05	COIL, 4.5 MHz trap & CTO
L-13	24D68517A19	COIL, 4.5 MHz audio take-off

INTEGRATED CIRCUITS

REF.	PART NO.	DESCRIPTION
IC-1	51D70177A02	INTEGRATED CIRCUIT: IF sound

DIODES

REF.	PART NO.	DESCRIPTION
D-1	48D67120A11	DIODE, low power
	48D67120A13	DIODE, low power (in panel BA-3 & later)
	48D67120A02	DIODE, low power (in panel BA-7 & later)
D-2	48C65837A02	DIODE, crystal
D-3	48C65837A02	DIODE, crystal
D-4	48S137133	DIODE, silicon: D3A

TRANSISTORS

REF.	PART NO.	DESCRIPTION
Q-1	48S137127	RF AGC DELAY: P2S
Q-2	48S134933	AGC AMP: A1V
	48S137171	AGC AMP: A6H (in panel BA-5 & later)
Q-3	48S134815	AGC KEYER: M4815 (USE 48-134910)
Q-4	48S137127	AUDIO DRIVER: P2S
Q-5	48S137169	AUDIO OUTPUT: A6G
Q-6	48S134981	1ST IF INTERSTAGE: A2Y (USE 48-134904)
Q-7	48S134932	2ND IF INTERSTAGE: A1U
Q-8	48S134932	3RD IF: A1U
Q-9	48S137172	1ST VIDEO: A6J
Q-10	48S137107	AUDIO AMP: A5M
Q-11	48S137168	AUDIO OUTPUT: P2V

TRANSFORMERS

REF.	PART NO.	DESCRIPTION
T-1	24D68501A17	1ST IF INTERSTAGE
T-2	24D68501A20	2ND IF INTERSTAGE
T-3	24D68588A04	3RD IF

CONTROLS

REF.	PART NO.	DESCRIPTION
R-14	18D66401A36	AGC SET-UP: 10K
R-27	18D66401A22	47.25 MHz TRAP ADJ: 100Ω
R-38	18D66401A22	4.25 MHz TRAP ADJ: 100Ω

MECHANICAL PARTS

REF.	PART NO.	DESCRIPTION
	29S10134A29	CONNECTOR, recept: panel mtg
	29S10134A32	LUG, terminal: AFC B+ take-off
	9C67349B04	RECEPTACLE, IF input

COMPLETE COLOR VIDEO PANEL "TA" OR "CA" CODED CA16 THRU CA24. Does not include "PA" Panel parts. See "PA" Panel Parts List.

REF.	PART NO.	DESCRIPTION
	1Y68662A59	COLOR VIDEO PANEL "CA" OR "TA" WITH "PA" PANEL: complete

ELECTRICAL PARTS

DIODES

REF.	PART NO.	DESCRIPTION
D-1	48G10346A02	DIODE

REF.	PART NO.	DESCRIPTION	REF.	PART NO.	DESCRIPTION
D-3	48S191A04	RECTIFIER, silicon (USE 48S191A07) (not in TA panel)	D-2	48G10346A02	DIODE
			TRANSISTORS		
D-4	48S137133	DIODE: D3A	Q-1	48S137172	1ST COLOR INT AMP: A6J
D-5	48G10346A01	DIODE (USE 48D67120A11)	Q-2	48S137172	2ND COLOR INT AMP: A6J
D-7	48G10346A01	DIODE (USE 48D67120A11)	Q-3	48S137172	AND GATE: A6J
D-8	48S137133	DIODE: D3A	Q-4	48S137172	AND GATE: A6J
D-9	48G10346A02	DIODE	Q-5	48S137021	DIODE, Zener: D1U
D-10	48S137133	DIODE: D3A		48S137337	DIODE, Zener: silicon; D5D (in panel PA-3 & later)
INTEGRATED CIRCUIT					
IC-1	51M70177A01	COLOR DEMODULATOR	**MECHANICAL PARTS**		
COILS & CHOKES				39S10184A31	CONTACT, card edge conn hsg
				26B68924A02	COVER, coil shield: panel mtg; L5, 7 & T1-"CA" panel
L-1	24D69708A02	COMPENSATING: 47 uh		15S10390A03	HOUSING, connector: card edge; 6 contact - less contacts
L-2	24D68852A03	DELAY LINE			
L-3	24D69708A02	COMPENSATING: 47 uh	**COLOR VIDEO PANEL "CA" & "TA" CODED "50" & "51" ONLY (SEE PAGE 68 FOR "CA-52" PARTS LIST)**		
L-4	24D68801A46	COMPENSATING: 800 uh		1Y68662A59	COLOR VIDEO PANEL "CA", "TA" CODED "50" & LATER: complete
L-5	24D68517A12	3.58 MHz OSCILLATOR			
L-6	24D68002A98	COMPENSATING: 5.6 uh	**ELECTRICAL PARTS**		
L-7	24D68517A21	HUE RANGE	**DIODES**		
L-8	24D68801A50	COMPENSATING: 47 uh	D-1	48G10346A02	DIODE
L-9	24D68801A48	COMPENSATING: 8.2 uh	D-2	48G10346A02	DIODE
L-10	24D68801A48	COMPENSATING: 8.2 uh	D-4	48S137133	DIODE, silicon: D3A
L-11	24D68801A48	COMPENSATING: 8.2 uh	D-5	48S191A04	RECTIFIER, silicon (not in TA panel)
L-12	24D68801A03	COMPENSATING: 100 uh	D-6	48S137133	DIODE, silicon: D3A
L-13	24D68801A02	COMPENSATING: 450 uh			
TRANSISTORS			**INTEGRATED CIRCUITS**		
Q-1	48S134841	1ST COLOR IF: M4841	IC-1	51M70177A01	COLOR DEMODULATOR
Q-2	48S134841	2ND COLOR IF: M4841	IC-2	51M70177A03	COLOR PROCESSOR
Q-3	48S134841	2ND VIDEO AMP: M4841			
Q-4	48S137002	VOLTAGE REGULATOR: A3M (not in TA panel)	**COILS & CHOKES**		
	48S137113	VOLTAGE REGULATOR: A5T (in CA-20 & later - not used in TA panel)	L-1	24D69708A02	COMPENSATING: 47 uh
			L-2	24D68852A03	DELAY LINE
Q-5	48S134842	PULSE LIMITER & INVERTER (USE 48S134992)	L-3	24D68501A25	OSCILLATOR: 3.58MHz
Q-6	48S137111	COLOR SYNC GATE & AMP: A5S	L-4	24D68801A30	COMPENSATING: 25 uh
Q-7	48S134842	CRYSTAL DRIVER: M4842 (USE 48S134992)	L-5	24D68801A30	COMPENSATING: 25 uh
			L-6	24D68801A48	COMPENSATING: 8.2 uh
Q-8	48S134841	CRYSTAL AMP: M4841	L-7	24D68801A48	COMPENSATING: 8.2 uh
Q-9	48S137115	ACC AMP: A5U	L-8	24D68801A48	COMPENSATING: 8.2 uh
Q-10	48S137003	CRYSTAL OSCILLATOR: A3N	L-9	24D66772A12	CHOKE: 7.5 uh
	48S137260	CRYSTAL OSCILLATOR: A7T (in TA or CA-24 & later)	**TRANSISTORS**		
			Q-1	48S134841	2ND VIDEO AMP: M4841
Q-11	48S137127	COLOR KILLER: P2S	Q-2	48S137127	SYNC & AGC TAKE OFF: P2S
Q-12	48S134970	PHASE SPLITTER: A2T	Q-3	48S137172	1ST COLOR INT AMP: A6J
Q-13	48S134841	PHASE SHIFTER: M4841	Q-4	48S137115	2ND COLOR INT AMP: A5U
Q-14	48S134918	3.58 LIMITER: A1L	Q-6	48S137127	"AND" GATE: P2S
Q-15	48S137113	BLUE VIDEO OUTPUT: A5T (in TA or CA-20 & later)	Q-7	48S137172	S-S SWITCH: A6J
			Q-8	48S137002	BLUE VIDEO OUTPUT: A3M
Q-16	48S137113	GREEN VIDEO OUTPUT: A5T (in TA or CA-20 & later)		48S137113	BLUE VIDEO OUTPUT: A5T
			Q-9	48S137002	GREEN VIDEO OUTPUT: A3M
Q-17	48S137113	RED VIDEO OUTPUT: A5T (in TA or CA-20 & later)		48S137113	GREEN VIDEO OUTPUT: A5T
			Q-10	48S137002	RED VIDEO OUTPUT: A3M
Q-18	48S137127	SYNC & AGC TAKE-OFF: P2S		48S137113	RED VIDEO OUTPUT: A5T
CONTROLS			Q-11	48S137002	VOLTAGE REGULATOR: A3M (not in TA panel)
R-76	18D66401A42	BLUE VIDEO DRIVE: 100Ω		48S137113	VOLTAGE REGULATOR: A5T (not in TA panel)
R-83	18D66401A41	GREEN VIDEO DRIVE: 100 Ω	Q-12	48S137172	HUE CORRECTOR: A6J
R-89	18D66401A40	RED VIDEO DRIVE: 100Ω	**CONTROLS**		
TRANSFORMERS			R-16	18D66401A49	COLOR KILLER: 10K
T-1	24D68517A18	2ND COLOR IF	R-54	18D66401A42	BLUE VIDEO DRIVE: 100Ω
CRYSTALS			R-55	18D66401A41	GREEN VIDEO DRIVE: 100Ω
Y-1	48C66865A04	3.58MHz CRYSTAL	R-56	18D66401A40	RED VIDEO DRIVE: 100Ω
			TRANSFORMERS		
MECHANICAL PARTS			T-1	24D68517A40	2ND BANDPASS
	26B66745A05	HEAT SINK, for transistor Q4, 15, 16 & 17	**CRYSTALS**		
	29S10134A29	LUG, connector: panel mtg	Y-1	48C66865A04	CRYSTAL: 3.58MHz
	43B68719A01	SPACER, transistor mtg; Q4, 15, 16 & 17	**MECHANICAL PARTS**		
INSTAMATIC COLOR PRESET PANEL "PA" (PART OF "TA" OR "CA" PANEL)				29S10134A29	CONNECTOR, recept: panel mtg
ELECTRICAL PARTS					
DIODES					
D-1	48G10346A02	DIODE			

REF.	PART NO.	DESCRIPTION
	26C66745A05	HEAT SINK: for transistors Q8, 9, 10 & 11 (A3M type)
	26C66745A07	HEAT SINK: for transistors Q8, 9, 10 & 11 (A5T type)
	3S136937	SCREW, tpg: 4-24 x 1/4: for transistors Q8, 9, 10 & 11 (A5T type)
	43B68719A01	SPACER, transistor mtg: Q8, 9, 10 & 11 (A3M type)

HORIZONTAL OUTPUT PANEL "DA-25" & "F-00 THRU F-25"

REF.	PART NO.	DESCRIPTION
	1U65534A89	HORIZONTAL OUTPUT PANEL "DA" & "F": complete
ELECTRICAL PARTS		
MISCELLANEOUS ELECTRICAL PARTS		
E-1	48S134921	DIODE, silicon: D1D
E-2	48S134921	DIODE, silicon: D1D
E-3	48S134959	DIODE, silicon: D1J (not in DA panel)
E-4	48S134959	DIODE, silicon: D1J (not in DA panel)
E-9	48D67120A13	DIODE, low power
COILS		
L-1	24D68778A01	HORIZONTAL OSCILLATOR
L-2	24D68801A16	COMPENSATING: 6600 uh
L-3	24D68573A01	HORIZONTAL SPOOK
L-4	24D69267A01	BALANCING COIL (in panel F-3A, F-5 & later)
	24D69267A02	BALANCING COIL (in panel F-9 & later)
	24D69267A03	BALANCING COIL (in panel F-13A, F-14 & later)
	24D69267A04	BALANCING COIL (in panel F-25 & later - not in DA panel)
TRANSISTORS		
Q-1	48S134917	DETECTOR: D1C; dual diode
Q-2	48S134842	HORIZONTAL OSCILLATOR: M4842
	48S137006	HORIZONTAL OSCILLATOR: A3S (in panel DA, F-6 & later)
Q-3	48S134910	HORIZONTAL PRE DRIVER: P1C
	48S134815	HORIZONTAL PRE DRIVER: M4815 (in panel DA, F-12 & later)
	48S137045	HORIZONTAL PRE DRIVER: P2G (in panel F-14A - replace with M4815)
Q-4	48S134919	HORIZONTAL DRIVER: A1M
	48S137093	HORIZONTAL DRIVER: A5F (in panel DA, F-23 & later)
Q-5	48S134910	ARC GATE: P1C (in panel F-0 thru F-8 only)
Q-6	48S134901	HORIZONTAL OUTPUT: A1D (USE A3H)
	48S134995	HORIZONTAL OUTPUT: A3H
	48S137203	HORIZONTAL OUTPUT: A6Z (in panel DA, F-25 & later)
Q-7	48S134901	HORIZONTAL OUTPUT: A1D (USE A3H)
	48S134995	HORIZONTAL OUTPUT: A3H (Q7 not used in panel DA, F-25 & later)
Q-8	48S134842	COLOR GATE DRIVER: M4842
CONTROLS		
R-23	18D68447A03	HORIZONTAL CENTERING: 10Ω (not used in DA panel)
TRANSFORMERS		
T-1	25D68782A01	HORIZONTAL DRIVER
	25D68782A05	HORIZONTAL DRIVER (in panel DA, F-13A, F-14 & later)
MECHANICAL PARTS		
	29S10134A29	CONNECTOR, recept: panel mtg
	76A544401	FERRITE BEAD: Q6 & Q7
	47C69019A01	POST, heat sink support
	9C63825A01	SOCKET, transistor: Q6 & Q7

PINCUSHION PANEL "GA"

REF.	PART NO.	DESCRIPTION
	1U65534A90	PINCUSHION PANEL "GA": complete
ELECTRICAL PARTS		
COILS		
L-1	24D68778A03	COIL, top tilt
TRANSISTORS		
Q-1	48S134838	VERTICAL PIN AMP: M4838
Q-2	48S137041	REG DRIVER: A4M
Q-3	48S137041	REG DRIVER: A4M
TRANSFORMERS		
T-1	25D68782A03	VERT PIN MOD.
CONTROLS		
R-2	18D68447A04	BOTTOM PIN AMP: 50Ω
R-9	18D66401A38	HORIZONTAL SIZE: 1K
R-16	18D66401A26	SIDE PIN: 100K
MECHANICAL PARTS		
	29S10134A33	CONNECTOR, recept: panel mtg
	26A66745A01	HEAT SINK: Q1
	43B68719A01	SPACER: Q1

CONVERGENCE PANEL "HA"

REF.	PART NO.	DESCRIPTION
	1Y68662A55	CONVERGENCE PANEL "HA": complete
ELECTRICAL PARTS		
DIODES		
D-1	48S191A08	RECTIFIER, silicon
D-2	48S191A08	RECTIFIER, silicon
D-3	48S10062A01	RECTIFIER, silicon (USE 48-191A08)
COILS & CHOKES		
L-1	24V68609A47	COIL, R/G right side vert lines: incls mtg nut
L-2	24D67682A08	COIL, R/G horiz lines
L-3	24D67682A03	COIL, blue horiz tilt (USE 24-67682A11)
L-4	24D67682A03	COIL, blue cent horiz phase (USE 24-67682A11)
CONTROLS		
R-1	18D67671A07	BLUE VERT TILT: 200Ω
R-2	18D67671A07	R/G VERT TILT: 200Ω
R-3	18D67671A05	R/G VERT DIFF AMP: 500Ω
R-4	18D67671A01	R/G VERT DIFF TILT: 30Ω
R-6	18D67671A01	BLUE VERT AMP: 30Ω
R-7	18D67671A09	R/G VERT AMP, 120Ω (some sets used 200Ω control) replace with 120Ω control and remove 270Ω resistor (R15)
R-9	18D67671A14	R/G - L.S. VERT LINES: 90Ω
R-12	18D67671A11	R/G HORIZ DIFF TILT: 150Ω
R-13	18D67671A04	BLUE HORIZ AMP: 150Ω
MECHANICAL PARTS		
	29S10134A29	CONNECTOR, recept: panel mtg
	29S10134A33	LUG, connector: jumper lead
	29S10134A32	LUG, terminal: jumper lead
	2C720979	NUT, coil mtg: (L1)
	9C67580A03	SOCKET, 12 pin

SWITCH MODE PANEL "JA"

REF.	PART NO.	DESCRIPTION
	1Y68665A32	SWITCH MODE PANEL "JA": complete
ELECTRICAL PARTS		
DIODES		
D-1	48S134921	DIODE, silicon: D1D
D-3	48S137021	DIODE, zener: D1U
D-4	48S134858	DIODE, silicon zener: M4858
D-5	48S10577A01	DIODE, silicon: 77A01
D-6	48D67120A13	DIODE, low power: blk-grn
D-7	48D67120A13	DIODE, low power: blk-grn
D-8	48S137347	DIODE, silicon: D5G
D-9	48S134939	DIODE, silicon: D1E
D-10	48S137347	DIODE, silicon: D5G
D-11	48S134921	DIODE, silicon: D1D
D-12	48S137348	DIODE, silicon: D5H
D-14	48S137347	DIODE, silicon: D5G
D-15	48S134939	DIODE, silicon: D1E
D-16	48S137347	DIODE, silicon: D5G
D-19	48D67120A13	DIODE, low power: blk-grn
COILS & CHOKES		
L-1	24D69267A03	COIL, balancing

REF.	PART NO.	DESCRIPTION
L-2	24D67534A24	CHOKE: 100 uh
L-3	24D67534A25	CHOKE: 430 uh
L-5	24D67534A23	CHOKE: 40 uh

TRANSISTORS

REF.	PART NO.	DESCRIPTION
Q-1	48S137281	THRYSTOR: W1L
Q-2	48S137093	DRIVER: A5F
Q-3	48S137172	SHAPER: A6J
Q-4	48S134943	REGULATOR: P1J
Q-6	48S137127	OSCILLATOR: P2S
Q-8	48S137341	OUTPUT: A8T

CONTROLS

REF.	PART NO.	DESCRIPTION
R-19	18D66401A38	OUTPUT VOLTAGE ADJUST: 1K

TRANSFORMERS

REF.	PART NO.	DESCRIPTION
T-1	25D68782A05	HORIZ. DRIVER
T-2	24D68778A04	SYNC
T-3	25D70783A01	SWITCH MODE

MECHANICAL PARTS

REF.	PART NO.	DESCRIPTION
	29S10134A29	CONNECTOR, recept: panel mtg
	14A562353	INSULATOR, mica: transistor socket Q8 (USE 14A543810)
	47C69019A01	POST, heat sink: Q8
	9C63825A01	SOCKET, transistor: Q8
	3S135232	SCREW, tpg: 6-20 x 7/16 clu pan hd; Q8

AFT PANEL "KA"

REF.	PART NO.	DESCRIPTION
	1Y68662A60	AFT PANEL "KA": complete

ELECTRICAL PARTS

COILS & CHOKES

REF.	PART NO.	DESCRIPTION
L-1	24D66772A12	CHOKE, RF
L-2	24D66772A12	CHOKE, RF

TRANSFORMERS

REF.	PART NO.	DESCRIPTION
T-1	24D68501A24	AFT DISCRIMINATOR

DIODES

REF.	PART NO.	DESCRIPTION
D-1	48S137299	GERMANIUM: D4R
D-2	48S137299	GERMANIUM: D4R

TRANSISTORS

REF.	PART NO.	DESCRIPTION
Q-1	48S134937	EMITTER FOLLOWER: A1Z
Q-2	48S134932	DISCRIMINATOR: A1U

MECHANICAL PARTS

REF.	PART NO.	DESCRIPTION
	42S10152A07	CLIP, push-on: AFT panel mtg
	15S10183A15	CONNECTOR, plug: 1 contact; less contact (AFT output)
	39S10184A02	CONTACT, plug: for 1 pin connector plug (AFT output)
	29S10134A15	LUG, connector: AFT input
	29S10134A33	LUG, connector: B+
	15S70096A01	SHIELD, AFT panel: less panel mtg clips

SYNC & VERTICAL DEFLECTION PANEL "VA"

REF.	PART NO.	DESCRIPTION
	1Y68665A33	SYNC & VERTICAL DEFLECTION PANEL "VA": complete

ELECTRICAL PARTS

DIODES

REF.	PART NO.	DESCRIPTION
D-1	48D67120A02	DIODE, low power
D-2	48D67120A11	DIODE, low power
D-3	48D67120A11	DIODE, low power

TRANSISTORS

REF.	PART NO.	DESCRIPTION
Q-1	48S137310	VERT. OUTPUT: P3U
Q-2	48S137309	VERT. OUTPUT: A8E
Q-3	48S137093	DRIVER: A5F
Q-4	48S137093	VERT. PRE DRIVER: A5F
Q-5	48S137127	SYNC SEP: P2S
Q-6	48S137172	SYNC INV: A6J

REF.	PART NO.	DESCRIPTION
Q-7	48S137172	VERT. OSCILLATOR: A6J

CONTROLS

REF.	PART NO.	DESCRIPTION
R-30	18D67678A13	VERT SIZE & LIN: Size 50K; Lin 4K

TRANSFORMERS

REF.	PART NO.	DESCRIPTION
T-1	25D67440A07	VERTICAL OSCILLATOR

MECHANICAL PARTS

REF.	PART NO.	DESCRIPTION
	29S10134A29	CONNECTOR, recept: panel mtg
	26C66745C07	HEAT SINK: Q3
	47C69019A01	POST, heat sink support (Q1, Q2)
	3S138066	SCREW, tpg: 3-48 x 5/16; Q1, Q2
	3S136937	SCREW, tpg: 4-24 x 1/4; Q3
	4S131147	WASHER, flat: metal; Q1, Q2

REMOTE PANEL "YA"

REF.	PART NO.	DESCRIPTION
	1Y68665A16	REMOTE PANEL "YA": complete

ELECTRICAL PARTS

DIODES

REF.	PART NO.	DESCRIPTION
D-1	48S191A02	RECTIFIER, silicon (USE 48S191A07)
D-2	48D67120A13	DIODE, low power
D-3	48D67120A13	DIODE, low power
D-4	48S137330	DIODE, silicon: D5B
D-5	48D67120A13	DIODE, low power
D-6	48D67120A13	DIODE, low power
D-7	48D67120A13	diode, low power
D-8	48D67120A13	DIODE, low power
D-9	48067120A13	DIODE, low power
IC-1	51M70177A07	VOLUME STEPPER

COILS & CHOKES

REF.	PART NO.	DESCRIPTION
L-1	24D68002A90	COMPENSATING: 6.6 uh

TRANSISTORS

REF.	PART NO.	DESCRIPTION
Q-1	48S134997	1ST PRE AMP: A3K
Q-2	48S137172	2ND PRE AMP: A6J
Q-3	48S137172	3RD PRE AMP: A6J
Q-4	48S137032	CHANNEL CHANGE PULSE DETECTOR: P2E
Q-5	48S137032	ON-OFF FUNCTION PULSE DETECTOR: P2E
Q-6	48S137172	PULSE SHAPER: A6J
Q-7	48S137172	PULSE SHAPER: A6J
Q-8	48S137127	ON-OFF RELAY DRIVER: P2S
Q-9	48S137171	SWITCH & LAMP CONTROL: A6H
Q-10	48S137171	SWITCH & LAMP CONTROL: A6H
Q-11	48S137343	LOW VOLUME SWITCH: W1P
Q-12	48S137343	MED. VOLUME SWITCH: W1P

RELAYS

REF.	PART NO.	DESCRIPTION
RE-1	80D69349A01	CHANNEL CHANGE
RE-2	80D69349A02	ON-OFF

CONTROLS

REF.	PART NO.	DESCRIPTION
R-10	18D66401A49	SENSITIVITY: 10K

COLOR VIDEO PANEL "CA" & "TA" CODED "52"

ELECTRICAL PARTS

DIODES

REF.	PART NO.	DESCRIPTION
D-1	48G10346A02	DIODE
D-2	48G10346A02	DIODE
D-4	48S137133	DIODE, silicon: D3A
D-5	48S191A04	RECTIFIER, silicon (not in TA panel)
D-6	48S137133	DIODE, silicon: D3A

INTEGRATED CIRCUITS

REF.	PART NO.	DESCRIPTION
IC-1	51M70177A01	COLOR DEMODULATOR
IC-2	51M70177A03	COLOR PROCESSOR

COILS AND CHOKES

REF.	PART NO.	DESCRIPTION
L-1	24D68801A30	COMPENSATING: 25uh
L-2	24D68852A06	DELAY LINE

REF.	PART NO.	DESCRIPTION
L-3	24D68501A25	OSCILLATOR: 3.58 MHz
L-4	24D68801A30	COMPENSATING: 25uh
L-5	24D68801A30	COMPENSATING: 25uh
L-6	24D68801A01	COMPENSATING: 15uh
L-7	24D68801A01	COMPENSATING: 15uh
L-8	24D68801A01	COMPENSATING: 15uh
L-9	24D68801A48	COMPENSATING: 8.2uh
L-10	24D66772A12	CHOKE: 7.5uh
L-11	24D66772A12	CHOKE: 7.5uh
L-12	24D66772A12	CHOKE: 7.5uh

TRANSISTORS

REF.	PART NO.	DESCRIPTION
Q-1	48S134841	2ND VIDEO AMP: M4841
Q-2	48S137127	SYNC & AGC TAKE OFF: P2S
Q-3	48S137171	1ST COLOR INT AMP: A6H
Q-4	48S137115	2ND COLOR INT AMP: A5U
Q-6	48S137127	"AND" GATE: P2S
Q-7	48S137172	S-S SWITCH: A6J
Q-8	48S137002	BLUE VIDEO OUTPUT: A3M
	48S137113	BLUE VIDEO OUTPUT: A5T
	48S137364	BLUE VIDEO OUTPUT: A8V
Q-9	48S137002	GREEN VIDEO OUTPUT: A3M
	48S137113	GREEN VIDEO OUTPUT: A5T
	48S137364	GREEN VIDEO OUTPUT: A8V
Q-10	48S137002	RED VIDEO OUTPUT: A3M
	48S137113	RED VIDEO OUTPUT: A5T
	48S137364	RED VIDEO OUTPUT: A8V
Q-11	48S137002	VOLTAGE REGULATOR: A3M (not in TA panel)
	48S137113	VOLTAGE REGULATOR: A5T (not in TA panel)

REF.	PART NO.	DESCRIPTION
	48S137364	VOLTAGE REGULATOR: A8V (not in TA panel)
Q-12	48S137172	HUE CORRECTOR: A6J

CONTROLS

REF.	PART NO.	DESCRIPTION
R-16	18D66401A49	COLOR KILLER: 10K
R-54	18D66401A42	BLUE VIDEO DRIVE: 100Ω
R-55	18D66401A41	GREEN VIDEO DRIVE: 100Ω
R-56	18D66401A40	RED VIDEO DRIVE: 100Ω
R-74	18D66401A47	NOISE LIMITER: 2.5K

TRANSFORMERS

REF.	PART NO.	DESCRIPTION
T-1	24D68517A40	2ND BANDPASS

CRYSTALS

REF.	PART NO.	DESCRIPTION
Y-1	48C66865A04	CRYSTAL: 3.58MHz

MECHANICAL PARTS

REF.	PART NO.	DESCRIPTION
	29S10134A29	CONNECTOR, recept: panel mtg
	26C66745A05	HEAT SINK: for transistor Q8, 9, 10 & 11 (A3M or A8V type)
	26C66745A07	HEAT SINK: for transistor Q8, 9, 10 & 11 (A5T type)
	3S136937	SCREW, tpg: 4-24 x 1/4; for transistor Q8, 9, 10 & 11 (A5T type)
	43B68719A01	SPACER, transistor mtg; Q8, 9, 10 & 11 (A3M or A8V type)

PART NUMBERS FOR COMPLETE REPLACEMENT PANEL KITS			
PANEL CODE	DESCRIPTION	PANEL KIT NUMBER	MOTOROLA REPLACEMENT NUMBER
A	UHF TUNER	KT220GM	1Y68663A99
BA	IF-AUDIO	KT194FM	1Y68662A57
CA	COLOR VIDEO	KT196FM	1Y68662A59
DA	HORIZONTAL OUTPUT	---	USE "F" PANEL 1U65534A89
F	HORIZONTAL OUTPUT	KT140DM	1U65534A89
GA	PINCUSHION	KT141DM	1U65534A90
HA	DYNAMIC CONVERGENCE	KT192FM	1U68662A55
JA	SWITCH MODE POWER SUPPLY	KT240HM	1Y68665A32
KA	AFT	KT197FM	1Y68662A60
TA	COLOR VIDEO	---	USE "CA" PANEL 1Y68662A59
VA	SYNC & VERTICAL DEFLECTION	KT241HM	1Y68665A33
YA	REMOTE	KT246HA	1Y68665A16